# Farming With Mary

# Giles Heron

Bright Pen

Visit us online at www.authorsonline.co.uk

A Bright Pen Book

Copyright text © Giles Heron 2009
Copyright graphics © David Moss 2009
www.mossarts.co.uk

Cover design by Mike Lyth ©

All rights reserved. No part of this publication may be reproduced, stored in a retrieval system, or transmitted in any form or by any means, electronic, mechanical, photocopy, recording or otherwise, without prior written permission of the copyright owner. Nor can it be circulated in any form of binding or cover other than that in which it is published and without similar condition including this condition being imposed on a subsequent purchaser.

The majority of personal names used in the text are fictitious.

ISBN 978 07552 1115 9

Authors OnLine Ltd
19 The Cinques
Gamlingay, Sandy
Bedfordshire SG19 3NU
England

This book is also available in e-book format, details of which are available at
www.authorsonline.co.uk

## Dedication

For all the volunteers, visitors and neighbours without whom we could not have restored Bank House

*Arthur
with happy memories
Giles & May.*

# Farming With Mary
## THE BANK HOUSE STORY

Front cover photograph:  Bank House Farm in distance, partly restored
Back cover photograph:  The Devon herd ('Red Rubies') at Bank House
Frontispiece:  The farmyard seen from the bank top

# Contents

INTRODUCTION — The Bank House Story

1. THE DECISION

Our personal background. How we came to give up a successful teaching career and buy a derelict hill farm.

THE FARM

The extraordinary variety and unique beauty of the farm despite its dreadful condition.

Map of the farm

2. PLANS AND PREPARATIONS

How we planned to use the farm in both human and agricultural respects.

3. THE GREAT MOVE

A graphic account of an unforgettable day : we couldn't have made it up.

4. UNDER WAY

Our first week in detail and other developments up to Christmas. Fifty visitors to stay including our first resident helper.

5. FENCING FENCING FENCING

Just what it says...

6. WINTERING CATTLE IN THE OPEN

A winter without buildings. Haying on open ground. Beginning to tame the heifers.

7. FARM SALES

Equipping the farm cheaply. Sales at the heart of rural community an essential education.

8. OUR FIRST CHRISTMAS — GEESE AND BOTTON VILLAGE

The blessings of kindred spirits at the nearby Camphill Village Community for the mentally handicapped.

9. THE SHEEP

A review of our first year's duties as shepherds and how we coped with them.

10. AT FULL STRETCH — OUR FIRST CALVING SEASON

Struggles and losses due to the poor fertility of the farm.

11. BLACK CARR BUILDINGS

The race to get barns and fold yards up in time for our second winter.

12  JEANNIE

Our first house cow: milking and butter making. A step towards self-sufficiency.

13  HOMAGE TO BONUS

Our initiation into bull keeping.

14  WOODLAND PORK

The virtues of pigs and keeping them in a nutwood.

15  HORSE POWER

Acquiring a Dales Pony. Training Krona to work with harrows, hay-turner and a sledge.

16  HAY HARVEST

The sociable climax to the year's work brings us closest to traditional rural life.

17  IN THE STEPS OF PIERS PLOWMAN

Learning to plough. The inescapable route back to fertility.

18  THRIFT AS A WAY OF LIFE

Extensive details of our domestic life: a pattern shaped by necessity and inclination: exploiting Nature's bounty to the full, recycling included.

19  A WORKFORCE OF VOLUNTEERS

The early spate of visitors and the growing sequence of long-term resident helpers: thumb-nail sketches and personal stories. Our life and work with them.

20  PROGRESS AND CHANGE

Improved crop yields and quality of livestock; sheep dogs; reclaiming further acres; stone walling and hedge laying; the National Park Tree Planting and Farm Management Schemes; oats instead of barley; tripods to the rescue; revived use of reaper and binder; free range poultry.

21  THE COMING OF AGE

More of our aspirations materialise; the most urgent work of restoration completed; volunteers harder to find in a Thatcherite climate as our energies decline. By a series of strokes of remarkable good fortune our successors are found and installed: graceful retirement secured for ourselves.

ACKNOWLEDGEMENTS

# INTRODUCTION — The Bank House Story

This is the story of the couple who bought a severely derelict hill farm in the North York Moors in 1972 with twin basic aims in mind. The first was to offer young people experience of voluntary, strenuous work out of doors through which they could earn the sense of purpose, achievement and self respect that in many cases their school days had failed to give them. The second, inextricable aim was to promote humane and conservationist practice in farming by showing that it is possible to win a modest but sufficient living from the land while foregoing many fashionable and particularly profitable systems of livestock management and uses of chemicals.

It is not the purpose of this book to argue what farmers ought or ought not to do with regard to these issues. Rather it is to point out that there are always alternatives to agricultural conventional wisdom and to record how we in our small way helped a considerable number of young people through an awkward stage in their lives by their participation in the revival of a moribund farm. It also relates how we shared the good life of animal husbandry with a much wider company of visitors as they took part in all sorts of farm work which they enjoyed enough to want to return time and again.

It was often the case that the extent to which our Bank House way of life was enjoyed and found helpful depended on its being experienced as a temporary liberation from the prevailing barren materialism in which most judgements about people were measured in terms of money and the ability to make it. In contrast we sought to be guided not so much by any idealism as by a true materialism, of the earth earthy, close to nature, beautiful and sustainable.

The book is designed to read like a story, fluently. It is moreover an adventure story with a strong element of risk and uncertainty throughout its series of leaps in the dark, for the survival of our venture was never guaranteed. As one lot of difficulties was overcome others appeared. Even so a sense of hopeful progress runs through the story as one aspect of dereliction after another was tackled. It is a 'warts and all' portrait that emerges; setbacks figure as openly as progress, mistakes as often as ingenious solutions.

The reality of its accounts owes much to lively quotations from contemporary letters and from the diary kept every day of the eighteen years we farmed Bank House. In spite of the lapse of some twenty to thirty-five years between events and their being written up here, there is written

# INTRODUCTION

evidence to support the veracity of everything recorded, which is hardly surprising coming from two Oxford historians. However, it is not so much this book's historical validity that warrants its publication as its relevance to the present state of society as we near the second decade of the twenty-first century. Social and economic conditions often circle round. There are strong similarities between the state of farms during the agricultural and rural depressions early in the last century and the state of Bank House Farm when we took it on. Similar too, alas, is the persisting failure to convince unacademic youth that it is more worth while striving to improve themselves and the community than relying on the National Lottery — and more enjoyable too. The forbidding prospects for dales farmers today revive memories of those earlier periods and increase the relevance of such stories as ours. Even more urgently, the dreadful present culture of knife-carrying urban youth should focus interest and attention on all positive ventures that have achieved any measure of success in aiming to provide youth with testing, educative experience of real life.

We do not represent what we did as any sort of blueprint for others to replicate — current Health and Safety Regulations alone rule that out — but it may encourage them to build on their own strengths and opportunities and see what they can do. The recent spate of radio programmes and articles in the press on these social and environmental subjects suggests this may be a favourable time for our story to be told more widely.

The Cottage, Glaisdale. August 2008.

Chapter One

# THE DECISION

One day in October 1972, in a train from Derby to St Pancras, we signed a cheque and sealed it in an envelope. It was our bid for a derelict hill farm in the North York Moors. Though we did not know it for some anxious weeks we had changed the direction of the whole of the rest of our lives. We were going to London in person because we were too late to trust our bid to the post. In order for it to beat the deadline we had to push it through a solicitor's letterbox that evening. We had left it to the eleventh hour, hesitating to abandon a secure and successful teaching career and to risk the entire savings of the ten years of our marriage. As the train rattled on we reviewed yet again the prudent cautions against the wild venture we were contemplating. Every line of argument ran into the buffers of impenetrable uncertainty. Had that neglected land degenerated beyond recall? Would our modest savings stretch to the purchase, re-equipment and re-stocking of the farm and still leave us enough to live on for the year or so before we could expect any income of substance? Did young people exist who would enjoy a period of strenuous, unpaid manual work in glorious country, and if so, were we likely to find them?

The rational case for accepting the challenge Bank House posed depended on encouraging answers to these and a host of other questions. Alas we had none of any certainty. In truth, to be or not to be was about to be determined, less by reasoning and sound advice than by instinct and blind faith. To this day we find it hard to understand how we had the nerve to reach the decision we did or indeed if it was really we that reached it. As at other key turning points of our lives – including our getting married – we felt more like performers in a drama already written than authors free to shape the plot. If the scene was to be set in Glaisdale, the name of the director was "chance".

The most influential factors affecting our future had been quite outside our control. In the first place we might have been blessed with our prime ambition, to raise our own family. That would probably have exhausted all our excess energy, time and money that in the event made our Bank House adventure feasible. As a childless couple we were free to take risks that would have been foolhardy for responsible parents. Alternatively, Giles might have been successful in one of various applications and interviews for headships and gone on to complete the full normal span of a career in education. Or the owners might not have decided to sell their farm at that time, or our estate agents might not have included the unusual notice of a 136 acre farm with "some rough" – a gross understatement – with dozens of others for bigger houses with about ten acres, which after all was what we had asked for. Or again, we might well have been dissuaded from making a bid at all by our best professional advice that auction under sealed tender was too much of a gamble for such derelict property at a time of worldwide economic crisis. Or of course we might have been outbid by someone with more money and an even greater disregard for business prudence.

We sometimes reflect that the odds against any event happening that does happen are always greater than those in favour. Certainly this seemed the case to us when, almost in disbelief, we signed the deposit cheque accompanying our formal bid for Bank House Farm. As we clattered on towards London we were seized by that sense of timeless unreality experienced on train journeys. In the symbolical dark of an early winter evening we were being carried away from our home and work in the comforting Dove valley, and even further from a remote farm on a hillside in North Yorkshire, to an alien metropolis which seemed to bear little relationship to either farm or school. Of such stuff dreams are made.

To say that we were contemplating giving up teaching for farming is so inadequate an explanation as to be distinctly misleading in regard to each half of the equation. Yes indeed Abbotsholme was a school and Giles had been a housemaster and Head of History there for thirteen years, for the last ten of which he had been Second Master as well. But Abbotsholme was more than just a school, and the job much more than just teaching. It provided education in the widest sense, for staff as well as for pupils, acquainting both with most interests and activities that genuine curiosity or original sin could devise. By turning hill farmers we stood to lose participation in a rich culture of music and drama, sport and hill walking, in addition to surrendering a secure salary, free use of a lovely Georgian house and garden and almost three months of holiday a year. In the event we were not to go abroad or sing in a choral society again for about twenty years. And of course we should be leaving behind a large, congenial company of colleagues, friends and neighbours inside school and out.

In two important respects the change in our way of life would be far from total. In the first place elements of farming had steadily infiltrated our school existence to the point of becoming Mary's major contribution. She had taken advantage of our one and a half acres of garden to raise geese regularly and to provide a lambing haven for a flock of ewes stranded on our doorstep by the foot and mouth epidemic of 1967. This resulted in the

# THE DECISION

establishment of the nucleus of a flock of our own whose descendants still flourish in Glaisdale. More important, Mary helped to retrieve the school farm from the brink of extinction, involving the pupils in the breeding and care of small numbers of cows, sheep and pigs, and in the use of George the carthorse. In Mary's experience of practical animal husbandry with Abbotsholme boys and girls, some of the seeds of our Bank House venture had been germinating for several years. The other main strand of continuity between school and the proposed full time farming was the educational purpose fundamental to the whole project. Our starting point was not a personal desire for adventure or just a change of scene for its own sake, attractive though these were, but our intimate knowledge of the common failure of most formal academic education to meet the deeper needs of many young people. We had a strong hunch that purposeful physical work, especially that concerned with the care of animals and the land, could be therapeutic much as spinach and wholemeal bread can help make up for the deficiency of a poor diet. We had known this prescription to succeed for many people, ourselves included. There was therefore never any question of our ceasing to work with and for young people. Rather we were searching for new ways of using our energies, talents and experience to complement the essential work of normal schooling. If in so doing we were able to assist other causes in which we were interested, such as the abolition of factory farming and a life-style based on maximum consumption of material goods, so much the better. The therapeutic, educational factor came foremost: the farming was to be a means to that end. All of which seemed very abstract at the time of our train journey. The day after tomorrow we should wake up back at school again. Bells would ring, pupils would clamour, lessons would resume and everything would be back to normal again except ......... except for the cheque book and its stub-end recording a big sum of money. What if it didn't come back?

Our search for a new life had been occupying us increasingly for some years accompanied by a degree of growing unease with the degree of privilege of our present existence. Could we pursue the educational principles we thought so valuable at Abbotsholme so that they were available more widely regardless of parental wealth?

Giles had reached that stage of a teaching career when the possibility of a headship could hardly be ignored. However, the prospect of becoming an administrator, almost a businessman, instead of the purveyor of interest and understanding, was distinctly unattractive. Moreover the discreet, self-effacing role of a headmaster's wife would not have come easily or naturally to someone of Mary's spontaneous and impatient unorthodoxy. We were

# THE DECISION

not interested in headship for its own sake, only if it offered scope for breaking new ground and a chance to resist a relentless demand for exam results. We were approached by Governors looking for a Head to establish a new ecumenical sixth-form college for Arab students in Jerusalem but decided it was not for us. An interview at a boarding school in a more extreme "progressive" tradition left it with the (correct) suspicion that Giles was insufficiently libertarian. Applications for two other schools failed for the opposite reason. One was for a mildly conventional Quaker school and the other for an old country town Grammar school with a single boarding house. These we had selected as being potentially suitable for a quixotic but serious scheme we had been planning, to run a boarding school with a residential home for old people, in the belief that segregation of young and old is to the detriment of both and that sharing activities and facilities would enrich the lives of both parties. Not surprisingly we found no buyers.

The more we looked at existing posts the clearer it became that if we wanted to put our ideas into practice we should have to create our own place and pattern of work. At this point, in the summer holidays of 1971, we chose to tour the border country and the Scottish Highlands with our car full of camping equipment and our heads full of ambitions. We were surprised to find how many big houses there were for sale and how moderate their asking price in some cases. After trespassing in the grounds of an eighteenth century mansion we learned from a nearby innkeeper that it was due to be pulled down because prolonged efforts to find a buyer had failed. We also learned that its last occupants had been a girls' school. That set us dreaming. Oddly enough really big houses seemed relatively cheaper than small ones. Might our small savings purchase a place in which to run one of our schemes? During the next twelve months we did a little arithmetic and a great deal more dreaming so that as soon as the summer holidays of 1972 began we set off on a repeat of our northern tour with the specific intention of tracking down some property inconveniently large for a typical family home but appropriate for whichever of our schemes we reckoned we could tackle in it.

To our surprise and disappointment the market had changed radically for the worse since the previous summer. We found that most of the empty houses we had seen had now been sold and the prices of all had risen sharply. By the time we reached Oban we realised that a housing boom was well under way which seriously threatened our ability to finance any of our projects. Furthermore, besides the rise in the prices of houses and land there was a wider economic crisis looming because of the disputes between the Arab oil producing countries. Inflation rose, share prices fell, making it an

## THE DECISION

unwise moment to sell our assets or to borrow heavily especially for so uncertain a project as ours. We realised that with the deteriorating economic situation we must buy our property very soon or postpone it for quite a long time. The casual approach we had followed so far, waiting like Mr Micawber for something to turn up, didn't promise well, so we began a more systematic search through estate agencies beginning, since that was where we happened to be, at Oban.

An estate agent close to the harbour found our enquiries so lacking in precision as to the size, character and price and location of the property we sought as to deter him from giving us notices of any appropriate pending sales. After that it seemed common sense to get in an open launch and pay for a trip to see the seals "money back if you don't see any". The seals were reassuringly real compared with our daydream property.

We extended our enquiries to England as soon as we returned home asking for a biggish unmodernised house with outbuildings and land for limited livestock, say ten acres. Through the early part of that autumn term we were inundated with advertisements for small farmsteads in mid-Wales which claimed that their greatest attraction was that recent motorway development had brought them within thirty to forty minutes of the centre of Birmingham. The extraordinary number of little farms for sale in the same area suggested that the prosperity and social cohesion of the district was threatened whereas we wanted to be part of a living agricultural community and not of a large scale invasion of Midlands businessmen.

Then out of the blue came the notice of the sale of Bank House Farm with its 136 acres in Glaisdale in the North York Moors, "some rough". Our first reaction was to pass it over without a second thought on the grounds that we could not afford so many acres nor could we imagine being able to manage them. What made us give it that second thought was those moors. We knew them to be as beautiful a district as any in England. We had spent our honeymoon with two of Giles' aunts who lived there in Lastingham and we knew how unspoilt it was in terms of its community as well as its scenery. A quick look at the map showed Glaisdale to be quite near. With so little else on offer might it not be worth having a look at it? If so, we should have to move quickly as there were only a few weeks before the bids had to be in. Unfortunately the following week-end Giles would be leading a hill walking party of pupils in the Lake District so Mary had to undertake a first reconnaissance on her own.

That solo trip made all the difference though not in any of the ways we expected. Having to find out of season bed and breakfast, she was directed

# THE DECISION

to a farm just up the dale from Bank House. Then she developed a bad cold which provoked the concern and generous cosseting of hosts which led to a greater intimacy and interest than a casual one night visitor would normally achieve. As a result Mary divulged the purpose of her visit and even outlined our plans. Instead of pouring cold water on the project as might have been expected the couple encouraged it and sent Mary on her way revived and remarkably well briefed as to the advantages and disadvantages facing any purchaser of Bank House. In a nutshell, the gist of her report back to Giles was that the farm was so seriously derelict in most respects, both the land and its buildings, that it was unlikely to attract either the competition or prices most farms were attracting at that time. Furthermore it was most unlikely to provide a reasonable income for quite some time. The normal advice to be expected from those best qualified to judge would obviously be not to touch it with a barge pole. On the other hand two factors in particular weighed more heavily with us than they might have done with other potential buyers: the exceptional beauty of the area and the quality of the farming community. Were these mere lightweight consideration in the scales against the enormous drawbacks or might they be sufficient to warrant a visit by the two of us together especially since time was running against us and we had no other immediate prospects to prefer?

The following week-end we booked in for bed and breakfast with the same, kind, potential neighbours and drove up for the briefest of visits. We beat the bounds of the farm and took photographs. The house we hardly saw for the vendors were unwelcoming and gloomy just like the photographs when printed, for the weather was dark and overcast. Then we had to dash back home armed with more than ample adverse evidence to begin, whenever school duties would allow, to assess, to discuss, to conjecture and eventually to decide.

We had fallen in love with Glaisdale, but it was not Glaisdale that was for sale, it was the derelict farm of Bank House. Wherever we had gone it was the same story of neglect and dilapidation: few internal field boundaries were stock-proof; tumbling stone walls had been stop-gapped crudely with barbed wire; many acres of pasture had been invaded by bracken, others by unchecked docks thistles and rushes; tracks had disintegrated and scores of thorn trees had established themselves on steeper pastures. It was also abundantly clear what a huge amount of drainage was desperately needed. Wasn't so much visible trouble enough to warn anyone off without the further hazard of invisible damage caused by neglected husbandry, especially the vital question of soil fertility which must have suffered from years of lack

## THE DECISION

of livestock? The only animals the vendors kept were a few pigs and Muscovy ducks.

Why then in view of all that we had found out and seen for ourselves was it not a foregone conclusion that we should sadly turn our backs on the Bank House sale as a glorious might-have-been and direct our search elsewhere? In some ways we shall never know or understand completely but we have some clues and there was not a total lack of method in our madness as our final debate in the train ended in signing our cheque and sealing our bid.

To begin with there was the setting. Dismal weather had not been able to hide the abiding beauty of the dale with the deep red of sodden winter bracken and its pattern of stone walls fretted across the dull November green of the valley bottom and the steepening slopes above. Who could settle in flat country knowing that such beauty had been on offer? More important however was the effect our encounter with Bank House had on our thinking. For too long we had been contemplating different projects, needing different settings, directed to providing different services, to plug various gaps in our welfare state. We had been unable to set estate agents identifiable targets because quite simply we did not know definitely which purpose they were to serve. Meeting Bank House and Glaisdale altered all that. Here was a specific challenge which narrowed our options to manageable proportions. From first glance it was clear that its isolation ruled out all consideration of care for the elderly as a main function. The limited accommodation similarly ruled out notions of a school as such and the most vague inclination we had entertained of commune life. What we were left with was essentially practical. Bank House cried out for restoration and the longer we considered it the more we realised how well that restoration could be fitted with many of our interests, our skills and our aspirations. Above all, we came to realise that the very serious degree of the farm's dereliction which frightened off conventional farmers and anyone for whom profit was a prime motive, was in fact an asset for us and a defining quality. There was so much to do in every way of unskilled hand work no-one could possibly be useless, no-one however backward or handicapped could fail to know he or she was contributing to a visible worthwhile improvement and helping to provide for our own necessities of living, growing our own food, fuelling our own fires and caring for our own livestock with whom the emotionally troubled if any such materialised could identify themselves.

Furthermore it was by no means all altruism. Our own selfish interests and longings would also be served. We had been increasingly at loggerheads with the crude materialistic values dominating social change; with ever rising expectations of creature comforts and intolerance or resentment if they were denied; with the suburbanisation of much of the countryside and the changes in agriculture itself. Here was a chance for us to live and work close to nature, close to the animals on whom we were to depend for a living and on a small scale in which the troubled would not find themselves insignificant.

Anyone still wondering at our final decision to sink all our savings and opt for a life of physical exertion and scanty spending power should consider that there are always people who are motivated by idealism and vision - more of them than are commonly recognised. In our eyes we were not buying dereliction that only too obviously existed, but our vision of what Bank House could become, and the prospect of the transition was itself most attractive. Right from the start we were buoyed up by the reaction of almost all our friends and relations who not only wished us well but wanted to share in our venture as far as their utterly different circumstances allowed. They were to prove an enormous support in the months and years ahead.

Verjuice Stone

# THE FARM

*God gave all men all earth to love,*
*But, since our hearts are small,*
*Ordained for each one spot should prove*
*Beloved over all.*

*Rudyard Kipling*

Glaisdale has surely been that spot for generations of inhabitants, native and incomer alike, and amongst them will have been those fortunate enough to have lived at Bank House Farm. Each of the red-roofed dwellings, strung like a necklace round the sides of the dale just below the spring line, is blessed with its own unique views of the features common to all. Below them, the tree-lined beck meanders from side to side of the dale through the flatter fields that provide farms with their most workable land. High above, some nine hundred feet above sea level, the apparently unbroken ring of the moorland horizon draws the eye like a magnet at every season, at every hour of daylight and often at night too. Linking beck and sky is an upward-sweeping curve of ever steeper pasture until the gradient becomes too sharp for horse or tractor, at which point grass gives way to rock, bracken, heather and bilberry.

Traditionally every farm needed access to both beck and moor, to water for cattle and to open moorland for sheep and peat, so the farmer's daily work often involved him in climbs of perhaps five hundred feet, an inescapable addition to the toil of whatever jobs were occupying him. North and south winds crossed high overhead from moor top to moor top, without dipping deep into the dale, but south-west and north-east winds have an awkward habit of funnelling the three mile length of the dale, gathering force where it narrows. Consequently the selection of task for any particular day usually took account of the direction of the wind, to take advantage of whatever shelter was afforded by the lie of the land, by the stone walls and plantations.

The above description could apply to the majority of dales in the North York Moors for they have so much in common. Glaisdale's unique character was given it at the end of the last ice age when the retreating, thawing ice deposited the moraine, blocking the mouth of the dale and creating a temporary lake which had to overflow somewhere. In time the overflowing water cut a deep, narrow ravine with walls still rising seventy or eighty feet on either side. It is the mound of the moraine which appears to complete the encirclement of the dale because the ravine is hidden by trees and by its own winding course. From most places in the dale it is not obvious where, or indeed whether, there is a way out for the beck. Anyone beating the bounds of Bank House Farm will discover this secret for himself because the beck below the

# THE FARM

eastern edge of the ravine forms part of the farm's boundary. The word "secret" is particularly apt since that end of the farm is largely wooded. Trees obscure the dramatic shape of the land as well as a number of archaeological treats in the forms of a shale tip, relic of attempts to find Whitby jet, a verjuice stone once used for crushing crab apples for the production of vinegar – an invaluable home-produced food preservative – and the mysterious foundations of a long deserted farmhouse, "Old Bank House". These last explain the presence of many crab-apple trees, a forty foot ancient dessert apple tree and several plum trees.

What makes that area a perfect playground for children, besides the spice of danger at the brink of the ravine, is the stream tumbling down the little valley it has made through a sequence of pools and waterfalls until it finally cascades twenty feet into the larger beck below. To complete the sensation of paradise, each spring a host of small, Wordsworthian daffodils appear in the dappled shade spreading from the beckside up the tributary valley and invading the edges of open pasture. No sooner have the golden heads begun to shrivel than an adjacent acre of woodland reveals its own magical carpet of bluebells.

All this is a world apart from the exposed rocky bluff arising most of three hundred feet immediately above the living farmhouse to which it gives its name, so steep it invites hands as well as sturdy legs for a direct assault on the crest which commands wide views of the whole dale ringed by moorland and of the village itself two miles off. Behind the rambler as he pauses for breath and admires the panorama, the top boundary wall conceals its own secret. Over seventy simple bee boles are built into it on the further side, recesses to house straw skeps – precursors of today's beehives – to harvest honey from the unbroken miles of heather. Away to the north-east, across a cleft in the hills, one can catch a glimpse of the sea's horizon beyond Whitby.

Two further treasures await inspection. The first is an unploughed meadow nestling within the meanders of the beck. Here thirty different species of flower are to be found blooming together in early summer, including meadow orchids. The other is a very steep eleven-acre wood, usually referred to as the "nutwood" because of the remains of hazel coppice, intermingling with the original mixture of hardwoods along with some rare ferns. The track which follows an easy contour to link the present Bank House with its older partner, plunges into the middle of the wood. Halfway through, on the lower side, wood suddenly gives way to steep pasture allowing views out across the dale from the sheltered walk in the manner of a cloister. Even now the whole has not been told. Can there be a richer variety of natural beauty within the compass of so small a farm?

KEY:
① Cattle Grid
② Farmhouse
③ Orchard
④ Black Carr Buildings
⑤ JCB's Grave
⑥ Footbridge
⑦ Old Bank House - remains
⑧ Vinegar Stone
⑨ Waterfall

⊞ ~ Bee Boles
π ~ Pig Houses
---- ~ Track
...... ~ Footpath

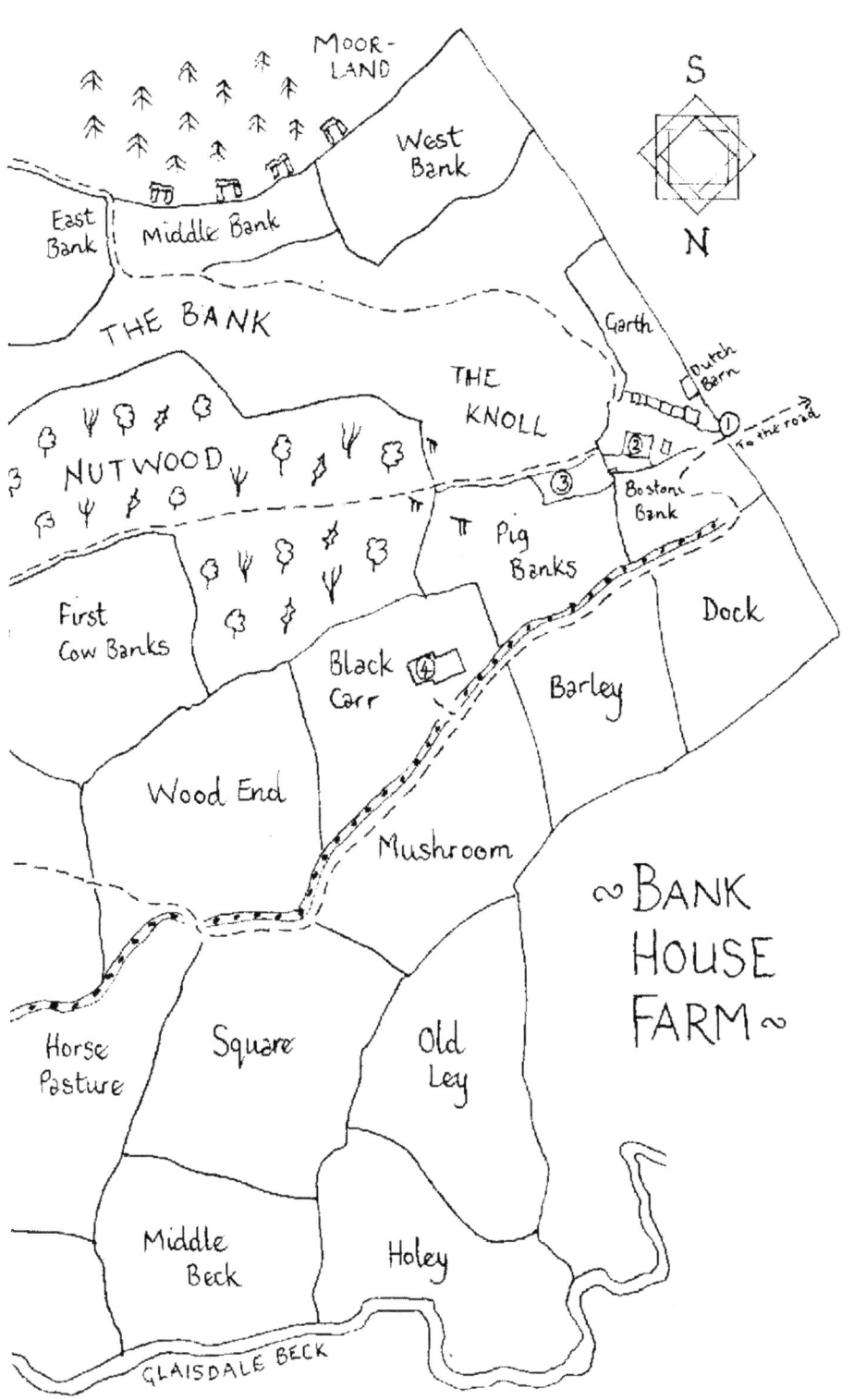

## Chapter Two
## PLANS & PREPARATIONS

It is even more difficult to keep a secret in a boarding school than in a small village especially if the matter to be kept secret is of intense general interest. All the more remarkable was it then that almost no-one besides the head master and ourselves had an inkling of what was coming when he broke our news to a weekly staff meeting. "The Herons have bought a derelict farm in North Yorkshire and will be leaving at the end of the school year." Stunned silence. The shock was twofold. That we should have got so far as to have arranged our departure without it being suspected or even rumoured was a cruel blow to deep-seated confidence in the prowess of the school's bush telegraph. Perhaps some of our earlier absences had been noticed, even commented upon, but since apparently nothing had come of them they had long been discounted and forgotten. It was only to be expected that at some distant day we should move on but the news that we had already bought a farm and were planning to revive it with our own hands, that took some believing. To pupils in particular we were fixtures in the landscape.

By agreeing to stay on till the following August we bought time for ourselves as well as for anyone else who stood to be affected by our going. It was appreciated that we were giving a generous period of notice in which successors could be appointed and prepared and our public exam candidates were particularly glad not to be changing horses or drivers, in mid-course. What was not so widely understood was how much we needed those nine months ourselves. To be more accurate it was Mary that had such a huge amount to do before we could start farming because the division of labour in the partnership we were planning was necessarily very lop-sided, at least until Giles had learnt enough to shoulder some real responsibility. For we were heavily reliant on Mary's past experience of running a collection of scattered holdings involving grassland management and small scale breeding of horses, pigs, dairy cows and, more recently sheep and geese. She had learned to deal with bankers, brokers and lawyers, with tax returns and accountants, and with all manner of officialdom. She and her mother had even defeated a determined ministry effort to build a nuclear power station on some fields the family owned on the Sussex coast.

This unusual combination of skills was promoted by the sudden early death of her father leaving her mother to bring up their six daughters of whom Mary was the eldest, the one with whom he had shared his work and aspirations most clearly. He had given up most of his practice as a barrister at Lincoln's Inn in 1939 and left London for the unspoilt lower slopes of the

## PLANS & PREPARATIONS

West Sussex downs, a move to be reflected much later by his daughter's move to the North Yorkshire Moors.

Giles on the other hand had had no equivalent preparation for a farming life. True, he had always liked using his hands and his physical strength and stamina. He remembered two days chaff carrying in 1939 trying to keep pace with an itinerant steam threshing machine; later he had learned to handle a scythe, a heavy axe and a cross-cut-saw; he was supple and nimble enough at potato picking not to delay the spinner as it circled round. However any farmer's son could do as much by the time he was eight, so it was no mere joke when he explained to incredulous visitors that Mary was the farmer and Giles the farm labourer. The one big area of responsibility allocated to Giles as we drew up our plan of campaign was that of negotiating with the young people we hoped would be attracted to join us, and directing their activity if and when they came. Of course every main issue was discussed and agreed between us, but in the early stages Giles's input was minimal and the importance of his involvement was essentially educational, including his own education.

Since at the time of our purchase Bank House Farm was not being farmed at all in any effective sense, everything had to be started from scratch, which gave us complete freedom to adopt the farming plan that integrated the unusual features of our scheme with all the common elements every farmer has to decide on: what to grow or breed, what manpower to employ, how to finance the whole venture. Our decisions on all these issues, and many others, had to be practicable and compatible. For example since our original motivation was to create suitable work for young people including some who needed the agricultural equivalent of a sheltered workshop, it was clear that most methods of conventional farming designed to maximise profit had to be severely modified or ruled out altogether. Whatever we produced and however we produced it had to aim instead for maximum job satisfaction and social cohesion. Much of the equipment we planned to use needed to be appropriate for a wide range of age, intelligence and experience, or lack of it. We did not wish to exclude someone because he or she had done nothing like farm work before, so we intended to rely on simple hand tools much of the time rather than on complicated, expensive, tractor-powered machinery, yet we could never get everything done with hand tools alone. We knew we should have to strike a balance and remain flexible enough to adapt to whatever personnel we assembled, which was in itself an entirely unpredictable factor.

## PLANS & PREPARATIONS

Another founding principle on which we were agreed from the start was that as far as possible everything we did should be in sympathy with the rapidly developing conservation movement which we preferred to think of as traditional farming. This was hardly surprising since before our marriage Mary had worked as personal assistant to Lady Eve Balfour at the headquarters of the Soil Association. In many ways this principle fitted admirably with our intended youthful workforce for whom chemical operations would be unsuitable to say the least. Furthermore the consequent dependence on hand tools for coping with weeds guaranteed purposeful work for unskilled teenagers. Similarly our repugnance of batteries and our determination to allow our poultry to range freely, would suit our personnel for whom the daily egg collection would be less of a drudge than a treasure hunt besides an opportunity to learn a great deal about the nature of hens.

We were not setting out to be bound rigidly to any doctrinaire organic rules. On the contrary, we were ready to compromise. For example, on the use of drugs we were prepared to treat individual sick animals but preferred not to follow the spreading practice of the general preventative dosing of whole healthy flocks or herds. As with veterinary drugs, so with weed killers: we did not intend to forswear their use altogether but to restrict it to treatment by spot spraying of individual weeds or patches of weeds instead of the blanket spraying of whole fields, again a policy compatible with our unskilled, young volunteer helpers. Gradually we put together a coherent body of the principles we hoped to be able to implement.

As our plans began to take shape we came to recognise that there was another general principle underlying most of our decisions and we were made conscious of it as we took part in the general debate then exercising people interested in rural life. Was farming or agriculture – the choice of the word indicated one's inclination – a business or a way of life? In a way it was a bogus debate because in most instances it was not a case of one or the other but of both. Nevertheless the question was interesting because answers given by a farmer told you a lot about him or her and about the farming community in which he or she operated. Family farms were more conscious of their way of life, whereas captains of the agricultural industry, each managing thousands of acres, emphasised the business element. For us the way of life took precedence. It was because we wanted to live in the way we envisaged that our Bank House project would make possible, that we would be farming at all and it was the way of life that we were wanting to share with others. Where adults were concerned it was sometimes described as an escape from the rat race. For the young however there was more than a whiff of adventure previously experienced only in camping: to spend most of the day in physical activity out of doors whatever the weather, a chance to wear whatever clothes you wanted, to keep irregular hours. It was to involve all sorts of new experience: the begetting and death of calves and lambs, piglets and poultry; the rounding up of sheep with or without a dog in order to dip them; learning to hang gates, to milk a cow by hand, to bake bread, lay a hedge or repair a dry stone wall. It was also a challenge to live cheaply, to survive especially in winter without many of the comforts increasingly expected in the later twentieth century. All the time we were preparing for this new life it began to seem less of a pipe dream, almost a reality. It was Mary's assignment that made progress; Giles remained fully engrossed in school responsibilities and the great question - how we were ever to find people to join us - remained a matter of faith almost to the very last minute.

As we moved into December there was one job we could not postpone. We simply had not time to write individual, different letters to each of our friends and relations, so, killing three birds with one stone, we composed a brief message of Christmas greetings which also served as advance notice of our change of address and gave all our friends as good an idea of what we were up to as we ourselves knew at the time. The message is shown below.

### MILLHOLME, ROCESTER 272, UTTOXETER, STAFFS
### CHRISTMAS 1972

*Stride over the moors from Goathland or Rosedale, catch a glimpse of the sea behind Whitby, scramble down into the green saucer of Glaisdale through the bracken and hazel wood to the red pantiles and low stone buildings of Bank House Farm, perched on a ledge above its lower fields that stretch along the near side of the beck.*

*We bought this 136 acre farm this November and shall move there permanently in August 1973, after the school year – our last at Abbotsholme – has been rounded off with a final Millholme music party*

*To begin we have our work cut out to rescue land and buildings from neglect. We hope to use them in a variety of ways, not just some practical farming to show that it needn't be all chemicals, batteries, and big finance, but also as a base to develop educational and therapeutic ventures, in fact a place and a home in which to help people.*

### BANK HOUSE FARM, GLAISDALE 297, WHITBY, YORKS

Born of necessity we had unwittingly started a tradition of sending out an annual news bulletin in place of Christmas cards. We assumed it was a "one-off" case of expedience. Every year since, however, we have

## PLANS & PREPARATIONS

considered returning to cards; every year since we are begged to continue with what we call our "Christmas letters"; and every year more of our friends and relatives retaliate by writing their own circular letters, much to our pleasure and benefit.

We had nine months between buying the farm in November and moving in the following August. Even while continuing to live – and work – over 160 miles away, there was a great deal we could do to ensure that when we finally took up residence we could hit the ground running.

First we had to be clear how we would farm Bank House. In many respects the pattern of farming we were to follow was determined many years earlier. From 1939 onwards Mary had been involved in the management of her family's land and livestock. This, together with her work at Haughley for the Soil Association, supported by many years of reading and debate, left her with a definite philosophy of man's responsibility towards the earth and its creatures, a philosophy in her case rooted in practical experience.

Until the unexpected advertisement for Bank House fell on our doormat we were not thinking of doing anything that would be described as farming, rather of simply "keeping some animals". But once fate had, so to speak, thrust it under our noses, Mary knew exactly what to do about it. It had taken her only two brief inspections of the farm and its state of dereliction to come to the verdict that the balance between the acres of potential hayfields or arable with those of rough pasture and woodland above should just enable us to make a go of it even on our limited funds and without abandoning our principles.

Most farms in Glaisdale had a dairy herd at that time, but dairying did not lend itself to our situation. In the first place we should have to borrow money for building and equipping a dairy itself. We had been ready to put all our savings into our unorthodox, risky venture but we were not prepared to add the burden of debt with little guarantee of any ability to repay. Then the exacting regime of dairy farming with its rigid and monotonous twice daily milking sessions, would not have been appropriate for our intended workforce. Furthermore, for some years to come most of each day would have to be devoted to repairing the damage wrought by the years of neglect. In any case since it was our wish to rear animals as naturally as possible we had no intention of following the common dairy practice of weaning or even slaughtering calves a few days old. With young helpers and other visitors in mind we wanted our cows to suckle their calves for many months. Our milk

## PLANS & PREPARATIONS

production was to be limited to one or two house cows milked by hand once a day, leaving the rest of the milk for their calves which they would be allowed to rear.

Just as we knew right from the start that dairying would play a very minor part in our farming so with arable work and for much the same reasons. Thus it took us little time to realise that the sheet anchor of our farming activity and income would have to be a suckler herd of beef cattle supported by a flock of sheep. The key to the return of Bank House to some measure of profitability had to be the improvement of its hay fields and the key to that was the restoration of grazing stock after an absence of at least nine years, reinforced by systematic application of the manure produced by the cattle in the winter half of the year. They would winter and calve in fold yards rather than in cubicles, not on expensive concrete but more cheaply on deep litter with fresh straw added daily. Thus instead of the more common daily chore of mucking out with tractors, we and our helpers would scatter straw by hand. Besides providing the cattle with warm, well drained bedding, this would ensure that the accumulation of the winter's muck remained under cover until Spring or early Summer instead of in open middens, so wastage and pollution were minimised. We should have to buy in a great deal of straw but in due course it would all find its way back onto the hay fields as muck, providing an organic alternative to artificial fertiliser. The higher ground would revive more slowly being grazed all year by a flock of sheep and by the cattle too in summer.

The existing buildings at Bank House were totally inadequate for a suckler herd, consisting as they did primarily of a low range of outbuildings entered by traditional split stable doors, far too narrow for any machinery mounted on a tractor. The only exceptions were a cart shed abutting the house and open to the east, its doors long vanished, and an exposed Dutch barn suitable only for storing hay and straw unless considerably adapted. The nearest approach to a milking parlour was some clumsy, stepped flooring in concrete, evidence of our predecessor's unfulfilled aspiration to keep some dairy cows some day.

To make it possible to keep our suckler herd as we hoped, by far our most ambitious and difficult job was to be the construction of a suitable complex of barns and covered fold yards. A great deal of time and effort was devoted to that end during our preparatory nine months. After inspecting and comparing many recent farm buildings in the neighbourhood we became very sure which firm we wanted to employ. Our preference was based on their use of timber throughout with a minimum of ironwork, on

## PLANS & PREPARATIONS

their high, well ventilated buildings with upper walls of Yorkshire boarding, also on their flexibility of design to meet our personal wishes and on the ingenuity of their accessories. After meeting them in January we designed the layout ourselves taking care to avoid it looking like a misplaced industrial hanger, creating instead the effect of a cluster, reminiscent of a hen with her chicks. We opted for grey-blue sheet roofing instead of the cheaper dazzling white that would take years of weathering to moderate its glare. Some years later national park policy made this a common condition of planning permission.

The most awkward problem proved to be its location. We wanted it to be in the lower fields near the middle of the farm, not an extension of the existing farmyard where there was no flat area large enough and which would involve the maximum use of the steep winding track for all hay coming up or muck going down. The Ministry, whose approval was needed for the large subsidy without which we could not have afforded adequate buildings, wanted us to put them on firm ground choosing of course an exposed position in one of our most valuable hay fields. We wanted to site them less obtrusively against the foot of the steep wood in a boggy field which we said we would drain. At that moment it was so wet the inspector merely looked over the gate from a safe distance and kept his shoes clean. Eventually he conceded to our choice provided our builder would confirm that it was possible, which the inspector clearly doubted. When we nervously invited the builder to come and vet the site he declined, asserting that he could, and would, build anywhere. He was based fifty miles away! Round one to the Herons.

While we were working on the new buildings to house our cattle we had to make up our minds on their breed and take steps to procure them. Almost every beef breed was considered, and then ruled out for one reason or another; continental breeds because Bank House could not produce enough fodder, Dexters produced too small a carcase, Highland and Galloways were too slow maturing, and so on. Mary spent many hours quizzing farmers about the suitability of different breeds for Bank House and for our particular needs. In December she was impressed by the Devons at the Smithfield Show and learned that they had won a competition for the best breed to cross with the ubiquitous Friesian dairy cows. However there were far more important reasons than rosettes for our selecting the Devons, affectionately known as "Red Rubies". They were exceptionally docile – even the bulls were manageable by their owners' daughters – and they had an established reputation for calving easily, two qualities of particular importance to us in view of the visitors and young helpers we hoped to attract. Other points in their favour for our farm were their relative immunity from bracken poisoning and their ability to thrive scavenging on indifferent grazing, since bracken flourished both within and without our boundaries, and much neglected Bank House land would remain poor for some time to come.

So Mary went to Exmoor to meet the secretary of the Devon Cattle Breeders' Society. He took her on a tour of farms with Devon heifers to sell and undertook to negotiate our purchase of eighteen pedigree heifers from four or five different herds to form the nucleus of our own future herd. They were to run with the bull in time to calve the following February or March, and a firm was engaged to collect and transport them to Glaisdale in August. In making our choice we had tried to be strictly business-like but once the die was cast we could not swear that the Devons' glorious glossy red coats and graceful horns hadn't been exerting a secret pull – a vision of red cattle in the verdant dale's landscape already enlivened with red pantiles and red bracken.

The early history of our flock of sheep was totally different from the way our Devon herd was established. Mary had been given a ewe lamb as a reward for having rescued a farmer from the dreadful predicament of being banned from access to his flock during the winter of 1967-8, including the whole of his lambing season, under the emergency foot and mouth regulations. Mary turned our garden into the equivalent of a temporary field hospital in which she acted as midwife to all his 50 ewes. That gifted lamb, fully justifying the name of "Winsome", proved a creature of remarkable character, health and intelligence. She was a born leader and an infallible weather prophet. She attended afternoon tea on the lawn, cropped the grass paths in the kitchen garden and knew exactly how long it was safe to switch her attack to our lettuces when Mary was distracted by the telephone. By the time we left Millholme in August 1973 Winsome was an established matriarch over five generations. In the course of eleven seasons she produced 21 of our healthiest and best grown lambs with prize-winning, fine Kent fleeces. For us to start with a tame, related band under its natural leader would be a great asset, offsetting the nervousness bound to affect animals uprooted from all their familiar landmarks, smells and sounds, especially helpful since at first we would be without a dog.

Our Devons helping to reduce the bracken

## PLANS & PREPARATIONS

Naturally we should have to add to this small family once it was established at Bank House and try different breeds of tups according to the demands of the Yorkshire market. But that lay in the future, and did not occupy us during this period as we gradually prepared for the move north.

Our sheep were not the only flock we had increased prior to moving from Millholme to Bank House. The policy that served so well with Winsome and her tribe could work equally well with Monsieur Mollet and his ladies. The task of preparing our geese for the great migration consisted mainly of setting more eggs in more nests in more hutches and remembering to shut them up at night. The net result was a cacophony of 16 shrieking geese ready to herald our arrival in Yorkshire.

Even with the good progress already described the agenda of the remaining necessary preparations was daunting. There was a limit to what we could do while we still lived well over a 300 mile round trip from Glaisdale especially while Giles remained fully occupied at school. Before we could move our cattle and sheep onto the farm we had to ensure they couldn't wander off it at will. We had to arrange with contractors to deal with the season's hay and arable programme until we were there to do it ourselves. We had to set in motion a major draining scheme which involved meshing the timetable of a Glaisdale drainage contractor with the wheels and maps of the bureaucrats authorising subsidies. To take advantage of a 70% local government re-roofing subsidy we had to find a builder to do the work. We also arranged for him to move the Aga to one side and construct an open hearth next to it. This would save money by enabling us to use the farm's abundant wood supply. We saved further money by rescuing three perfectly sound box sash windows from the demolition team's bulldozer in Uttoxeter and brought them north in our sheep trailer. The only cost was a ten-shilling tip for a workman. To avoid leaving the house unoccupied and damp all winter we had to contrive a legal transaction for buying the farm in such a way that the vendor could continue in residence until Easter without our risking his acquiring squatter's or tenant's rights to stay longer. For the sake of health and economy we had to establish a kitchen garden at the farm capable of keeping our household in vegetables and fruit for much of the year. For Mary in particular it would be a valuable saving in shopping time and petrol if we always had home grown food on the premises.

Realising that when we eventually moved in we should have very limited time to devote to house and garden as distinct from the farm, we determined to make the most of our last school holidays. Camping in the empty farm house for a week in early April was rather like a dress rehearsal

## PLANS & PREPARATIONS

for the rigorous life to which we had committed ourselves. The house was thoroughly damp and cold. Camp beds and bales of straw were our only furniture. The water supply ran out each midday and evening and a storm drain in the yard was blocked. However we were able to light wood fires, and once we had learned to manage the Aga we even achieved the luxury of a hot bath, albeit without any cold water. The weather proved particularly educative with gales, hail, sleet and snow teaching us the hard way how much later and fiercer spring can be in the north. Nevertheless we laid the foundations of gardens on each side of the house. The list of what we planted tells its own story. In front, on the south side, to the few surviving roses we added Solomon's seal, St John's wort, dahlias, delphiniums and lupins, chrysanthemums, aquilegia, and phlox, honesty, gladioli, mullein and cranesbill, dondia and evening primroses. Clearly among the many aspects of civilisation we were prepared to deny ourselves in our new life the joy of the colours and scents of flowers was not to be one.

On the north side of the house we began to convert what was called the peat yard into a fruit and vegetable garden. Here we planted sprouts, cabbages, spinach and onions, artichokes, carrots and rhubarb, peppermint, potatoes and parsley as well as some gooseberry bushes. More was to come with us in August.

In that chill April week when the weather did its worst we were glad to make progress indoors as well. We hired a heavy sanding machine and choked ourselves with sawdust as we improved the floors of sitting room, staircase, landing and bedrooms, having to overcome the unexpected handicap of iron tacks. The tedium of their removal was turned into a game by keeping the score: the top step yielded just over a hundred!

This prologue to life at Bank House turned out to be more sociable than we had expected. Giles's sister and her two youngest children joined us for two days entering into the spirit of adventure whole heartedly and most helpfully. We spent quite a lot of time discussing work with a local builder so that he could get on with it during our absence. Our predecessor called to retrieve a knife sharpener. Then the ministry vet came to see us about brucellosis testing though we did not yet have a cow on the farm. Next we toured the farm with the ministry drainage advisor looking for a suitable source of water for the proposed cattle yards. His jug and stopwatch established that one quart per second – about ninety gallons an hour – was pouring down the hill above the boggy field on which we hoped to build our fold yards. He assured us this would be ample. We met the vicar in church, complete with skull cap, and also made contact with the postman, the dustbin man and the milkman. We even found time to commune with the wild daffodils above the ravine before returning for our last term at Abbotsholme.

Life at Bank House was fast becoming fact rather than fiction, its reality emphasised by the continuous process of acquiring the equipment that we expected to use, agricultural and domestic. Of course we should go on looking for things once we were there but the financial inducement to buy second hand wherever possible led to a haphazard order of acquisition. And since a wide circle of friends and relations wanted to assist our venture we received a great many gifts, often of things not on any of our lists. For instance a salt of the earth Staffordshire pig keeper's widow said we would have to have lots of floor cloths and proceeded to hem very serviceable items made from discarded Lyle stockings, winter underwear and sacking. She also accompanied Mary on a flying working visit to the farm in June, saying that Mary would get more done if she didn't have to cook and wash up. She helped prepare more vegetable beds too. Blessings on her memory and on many other generous souls!

At the other extreme of the range of equipment needed were a sheep trailer and a tractor. We weren't sure we could afford a Landrover and planned to make do with our old estate car if we could fit it up with a second-hand trailer. In terms of value for money we never did better than with the old sheep trailer which proved invaluable ferrying loads of our belongings to Glaisdale before its years of service with pigs and sheep.

The matter of a tractor was much more serious and problematic, involving research and courage on Mary's part. She was advised that what would best suit our particular needs and terrain at Bank House would be a Massey Ferguson 135 but she learned that they were so popular in the dales that they were extremely hard to come by in Yorkshire and dealers had long waiting lists. So when she saw a Staffordshire sale notice advertising one "almost new" it was a chance too good to miss and she set off in determined mood. There were many bidders and the competition was fierce. When the auctioneer finally knocked it down approaching, but still below, its show room price, the ring of cloth-caps and gaiters almost broke into applause for the little lady hitherto known only for buying old arm chairs for school playrooms. Which explains how Giles found himself monarch of all he surveyed over the hedge tops, driving a tractor for the first time in his life without any proper instruction as he learned the hard way in narrow lanes what is the length and breadth and height and weight and braking power of a tractor. His pride of ownership was at loggerheads with his obvious

## PLANS & PREPARATIONS

incompetence. It was the sight of the gleaming red tractor parked in the Millholme garage that first brought it home to Abbotsholme pupils that this story of our farming was for real. It brought it home to Giles with similar force. Still oddly shy about that incompetence he sneaked off with the school farm manager when pupils were safely in class to watch the hay cutting and even to take control for a few clumsy circuits, the briefest of apprenticeships but an invaluable one.

Throughout these last months we kept our eyes and ears open wherever we went, arriving home with booty from the garages, attics and kitchen cupboards of friends and relatives, with spare beds and blankets, household and garden tools, large casseroles, unwanted buckets and candlesticks. An antiques shop in Reeth yielded wide, shallow steel pans for setting milk, a farm sale an infolding cast iron mangle since we should have to do our own laundry in future, and from somewhere else a glass, table top butter churn and butter pats. To cap it all as the school year ended we were showered with practical leaving presents, a Black & Decker power drill, a sockets spanner set, a lantern, a knapsack and even a dark red nine foot five barred gate.

We also resorted to barter. The only dining table that we possessed large and strong enough, doubled up as a billiard table to which the Millholme boys had become much endeared. To drown the chorus of woe at the prospect of losing their billiard table we did a deal with the bursar and swapped it for a large oak trestle table used for sorting school laundry. It remains our dining table to this day.

By the end of our nine months' preparation we had provided for almost all of our foreseeable domestic needs. What we still lacked was mostly of the nature of agricultural tools and machinery that could most conveniently be obtained at local farm sales in Yorkshire once we were there ourselves. Before that however we had other hurdles to overcome. One, how to get everything shifted to Yorkshire, had long loomed ahead and we were ready for it with a detailed plan of campaign. Another however took us by surprise and demanded an immediate, improvised solution.

For some months we had felt secure about the purchase of our Devon heifers, assuming we needn't concern ourselves about their delivery until we were in Glaisdale to receive them. The longer they ran with the bull the more likely they were to be in calf. It was a blow therefore to learn that if we didn't arrange for their transference to Yorkshire by July 10th there would be a substantial increase in the bill in the form of a charge for their further keep. It would have been simplest just to sell some remaining stocks and shares and pay the extra, but the Arab oil crisis had caused a slump on the stock exchange so it was the worst possible time to sell shares. Instead we arranged for the heifers to be delivered to Bank House on July 10$^{th}$ and we asked our Glaisdale neighbours to keep an eye on them for a month until we moved in. This was not much of a burden for them as they were already going over to our fields most days to see their own cattle which were there under an agreement we had made with them at Easter. They could enjoy free grazing of some 25 acres in return for their doing whatever fencing they found necessary to prevent their beasts straying. However we couldn't prevent our own unsupervised heifers straying without our doing some immediate fencing repairs ourselves. So Mary made a previously unscheduled week-long visit to the farm. Fortunately the summer term was almost over and many pupils were at a bit of a loose end having finished their public exams so Mary was able to take four boys with her to create a temporary safe zone for the heifers. Giles managed to reinforce the squad at the weekend, taking a couple of girl pupils as well. Only Mary was free to stay on long enough to supervise the heifers' arrival and witness the stir made by these unfamiliar red horned cattle amongst the local farmers. After that, with fingers crossed, she had to leave them to their own devices and return south, first for the celebrations that wound up the Abbotsholme school year and then to prepare for the last of our annual music weeks at Millholme. We filled all the boys' vacant beds with the friends we assembled to form a choir to sing under the tuition of a first rate conductor. When we were not singing, dancing on the lawn, peeling potatoes or attending to the sheep and geese, we would walk along the banks of the river Dove or play croquet. Only when we had recovered from the final all night open air party, could we face the last hurdle, the long-awaited removal from Millholme to Bank House.

## Chapter Three

## THE GREAT MOVE

Flitting is a term commonly used for moving house and home, though whoever first coined the term may have had his tongue in his cheek. The speed and ease of movement which the word suggests could hardly have been further from the reality of our experience when moving to the North York Moors. The months of planning that went into it resembled a complicated army manoeuvre rather than a swallow swooping in and out of a barn.

Every removal will present its own particular difficulties and ours certainly promised us our share of which the least was the basic packing of household possessions and loading them into a furniture van. In addition we had to arrange for the transport of a tractor, a heavy old farm trailer (a recent gift), sixteen ten-foot five-barred gates and nine 12-foot ditto (bought at a discount, courtesy of the local hunt), eighteen geese, sixteen sheep and the fencing and posts to contain them.

We had to work to a tight schedule because so much seemed to need to be left to the last minute. Since fortune favoured us with a heat wave after weeks of rain, priority had to be given to the sheep and geese so that their time in the inescapably sweltering cattle van would be as short as possible, but the job of taking down the posts and rolling up the wire netting had to wait until the animals were already on board.

Another claim to a late place in the batting order was a large, heavy freezer packed choc-a-bloc with a year's garden produce which was to tide us over till the new Bank House vegetable garden, planted after Easter, could replenish it. How long could it avoid thawing under the fierce August sun once it was disconnected from the Millholme electricity supply? Then again, the survival of the large number of plants we were determined to take with us, depended on their not being left drying out too long before being re-planted. If it hadn't been for the unusual heat we might have been able to dig them up a day or two earlier.

Yet another job that couldn't have been completed in advance was changing the plugs on every electrical appliance; it was remarkable how many there were! Millholme had been fitted with square-pinned, 13 amp power points and plugs whereas, almost symbolically, Bank House remained equipped with ancient, brown bakelite round pinned, 5 and 15 amp power points. It had proved very difficult to find enough discarded round pin

# THE GREAT MOVE

plugs so that we could arrive at the farm able to boil a kettle, turn on the television or read in bed. Each appliance had to have its plug changed before it could be used again.

As we considered these rival claims to be left to the end we realised there was yet one more with the greatest claim of all, Tacitus our cat and her pair of suckling kittens. This cannot be fairly levelled at us as a failure in planning, family or otherwise. Not long before, a kind villager had walked into Millholme desperately seeking a home for a kitten found deliberately shut in a neighbour's cupboard. Since our old cat had disappeared without trace and we were assured that the little charmer on offer was a tom, we had accepted it as a useful addition to our future farm menagerie. By the time the misinformation was proven it was too late to avoid the most awkward of our removal problems, how to avoid the escape of kittens or cat from a house full of strangers and disturbing noises, or from a car once we had hit the road.

Such were our most obvious problems. The strategy we had adopted might have worked smoothly, with a bit of luck. Clearly a single conventional removal van would not suffice. We chose to hire a furniture van from York thinking that a Midland driver might not believe how much steeper the North York Moors roads were than those round Millholme, as we had found to our cost when we had towed our first trailer load at Easter, arriving in Glaisdale by a sequence of gradients, 1 in 3, 1 in 4, 1 in 6, 1 in 4, and 1 in 3, again. The final shortcut we took on which we had actually got stuck wasn't even given a listing.

We arranged for the furniture van team from York to arrive at 9 a.m. and to spend the morning loading up with everything in the house. They were to spend the night at York and reach the farm next morning where we should be ready to receive them. So far so good. We also engaged a cattle van from Glaisdale itself. It was to arrive at noon by which time we hoped to be able to attend to loading it with the tractor, and then the sheep and geese and their fencing. Since we calculated there wouldn't be room for the gates and the heavy trailer as well we had, with some misgivings, accepted an offer from an old Staffordshire farmer and Methodist preacher – his farm was minute and he always wore a bowler hat – to get a friend of his with a 50 foot Volvo to bring everything else "before long" for the cost of the petrol only. Oddly enough this last arrangement was to prove the most satisfactory of the three.

To complete our transport: once the cattle van had left, Mary was to drive "Calvi" – our open Morris Minor of a light blue that reminded us of

## THE GREAT MOVE

the Corsican bay where Nelson lost his eye – fast enough to overtake the van and reach Bank House before it. Giles was to wind up the proceedings in our estate car "Trafalgar", the name on the manufacturer's colour chart of its tint of deep blue. Trafalgar would tow our sheep trailer laden with the heavy freezer. The car itself would carry lots of plants, the cat and her kittens and a bag of 5 and 15 amp round pin plugs, together with an electrician's screw driver.

It was very late before we staggered to bed for our last night at Millholme, our heads full of all the above arrangements and of the 101 other things we still had to do in the morning before the van from York was due to arrive at 9 a.m. Our peace was ended unduly early by the vibrations of a heavy vehicle in the road outside just before 8 o'clock. Blast it! Irritation turned to utter dismay when we discovered that the noise heralded not the expected furniture van from York but the Glaisdale cattle van. The driver manoeuvred into the yard and explained airily that he was over four hours early because he now had another job back at home that afternoon. He seemed to expect us to help him load up straight away. Trying to repress our wrath we said we would help him when we were free but we would have to leave him to it and attend to the furniture removal team when it arrived and also to Mary's sister and her vicar husband who had taken the day off to dig up all the treasured plants we were taking to Glaisdale, some of which had begun life in Yorkshire many years before in relatives' gardens. Of course the plot thickened once the York van arrived, more or less on time. We had a frantically chaotic morning skipping between the different operations; explaining which rooms were to be left untouched – those with school property and those with the assistant house master's possessions; marking particular plants to be dug up; pinning a notice on a door – "Young kittens! Do not let out!"; trying to coax sheep and geese to take on board more water than usual since their next refreshment would be inexplicably far off; making and distributing coffee at regular intervals in the few unpacked mugs; finding one set of tools to help the furniture team to dismantle part of an awkwardly large wardrobe and another for the Glaisdale man to take the roof off the tractor. In between there were rare moments in which to complete our personal packing.

The tractor gave us a fortunate breather by occupying the driver for quite a time. Although we had sent him precise, and as it turned out accurate measurements, nevertheless when he drove it up the ramp the van was too low. First he let the tractor tires down. When that wasn't sufficient he tried to remove the cab roof and dismantled the exhaust pipe. This unexpected complication took him off to the village garage to borrow and

# THE GREAT MOVE

later return effective tools. Our grip on the proceedings was rescued from probable melt down by the unplanned, spontaneous contribution of two more pairs of hands, those of the headmaster and his wife. They had already spent many hours in the preceding days packing all the contents of kitchen cupboards and dressers. They had an uncanny knack of knowing when and where their help would be most welcome, when an extra heave would shift the immovable and where a brush or a bag of string was last seen. While we were securing the geese and sheep, first in the inner yard and then in the van, they were already rolling up the fencing and making neat parcels of the posts.

In spite of all the obstacles the cattle van drove off soon after lunch, or rather, soon after a latish lunch time, having severely disrupted our timetable. Mary had to rush through her remaining jobs, gobble some of the welcome picnic lunch thoughtfully produced by well-wishers, hold final councils of war about how to pack the plants, and set off in hot pursuit of the Glaisdale van in the Morris Minor, mercifully with its roof down.

At this time the pace slackened allowing fatigue to extend the picnic on the lawn. Giles returned to supervision of the furniture removal gang who had done sterling work until the very last moment. Now however their patience wilted in the heat just as they came to what could well be called the pièce de résistance, a family antique, the most valuable and delicate item we possessed. It was a six foot cheval-glass complete with finials and marquetry. To protect it from the hurly-burly it had been put in a separate room and left to the end. Suddenly the boss appeared and said they were off. "What about the tilting mirror?" The boss looked uncomfortable. "Hang on!" Giles dashed upstairs, grabbed the glass which had been detached from its fragile stand and carried it out into the street where the men were already closing the back doors. Reluctantly they reopened the doors and looked for some rope to secure the glass. "Hang on!" – again. Giles pelted back upstairs and then struggled down more cautiously, with the awkward frame and out into the empty street. The van had gone!

By now Trafalgar itself was ready for departure. The sheep trailer, weighed down by the freezer, had been hitched on behind like a gun carriage. The car itself looked like a cross between a hearse and a florist's delivery vehicle crammed with open boxes of plants many of them in full flower, brushing the roof. The incongruous forlorn antique stood by its side out on the lawn in the sun. Could it possibly squeeze in over the plants? Very gently we eased it horizontally through the flowers and poked its finials over the top of the front passenger seat. Its branched feet protruded through

# THE GREAT MOVE

the open hatchback. No way could the hatchback be closed though it seemed unlikely that anything would fall out. But what about Tacitus and the kittens?

We had made a comfortable bed for the kittens where the front passenger's feet would have been and by its side a freshly filled earth box. The passenger's seat itself was naively allocated to Tacitus, that is if she didn't prefer the driver's lap. Nothing remained to do but retrieve the little family from the safe custody where they had been shut up throughout the long day's drama. The kittens posed no problem at first as the car made a welcome change of scene after their confinement. Tacitus' maternal instincts and exploring propensity made it possible to get her to enter the car and rejoin her offspring. There cooperation ceased. The instant Giles started up the engine and pulled gingerly into the street, Tacitus set up a wail which she kept up for the first twenty miles, pacing backwards and forwards incessantly by the windscreen, summoning the kittens to join her. Giles could do nothing to control Tacitus. He had his work cut out driving with one hand while keeping the kittens down in their quarters with the other. Every time one scrambled up onto the seat he grabbed it and thrust it back into its bed. This game continued until the motorway was reached, by which time the kittens conceded defeat and fell asleep. Tacitus too changed tack and began to explore the jungle of plants, weaving in and out, still complaining from time to time. Once we were speeding steadily up the motorway she discovered the open hatch. One moment Giles saw her in the driver's mirror. She was gazing fascinated by the road flashing away beneath her. The next time he glanced up, she had vanished. After that there was neither sight nor sound to indicate where she was.

While Giles was eating up the miles on the M1 he was imagining his arrival at the farm and rehearsing his explanation for the absence of Tacitus and why the kittens were probably orphans. He was also imagining, with a tinge of envy, that Mary would have reached Bank House and, having brewed some tea, would have taken it out to enjoy the evening sun sinking over Glaisdale. However she was not in fact as far on her way as he assumed for she had drama of her own to contend with.

In those days a car journey from Millholme to Glaisdale was likely to take the best part of four hours even without a trailer. The York and Malton bi-passes were not complete so the direct route lay straight through the middle of both towns which in York took one along the south side of the Minster. Just after that the road narrows a great deal to squeeze through Monkgate, the medieval gateway in the city walls. Exactly there, under a

## THE GREAT MOVE

threatening portcullis, Mary's journey came to an abrupt halt. Calvi had a puncture. It was the height of the rush hour and the spare wheel was buried under loose belongings that could hardly be said to have been "packed", merely "packed in".

Instantly the traffic piled up behind her and Mary was besieged by a crowd of irate motorists. She decided to play the helpless female. Thereupon the motorists took the law into their own hands and a group of the strongest picked up Calvi and deposited it on the nearest pavement, outside a bicycle shop. Possibly doubting the legality of their action they negotiated with the staff of the cycle shop to help Mary change the wheel before melting away. It was not much longer before Mary too was on her way once more.

As she finally drove into the Bank House yard she was disappointed, though hardly surprised, to see no sign of the cattle van. The yard was desolate and empty except for the geese queuing up by the stable door (which some intuition told them might offer suitable protection) and the sheep surveying their new kingdom from the upper corner of the yard. The latter looked disappointed by their immediate prospects and distinctly reproachful, however they had obviously found the water troughs and anyway it was at last getting cooler. Mary turned to go indoors..

The key worked, and so did a kettle and the emergency provisions left over from the last visit. She looked out of the kitchen window to see how the leeks, sprouts, onions and broccoli were faring that she had taken so much trouble to plant. Horror of horrors! The back garden looked like a battle field: short stumps scattered everywhere were all that were left this side of some trampled fencing. The culprits, our own red heifers were nowhere to be seen. Who would not have wept? Mary, exhausted as she was, certainly did. Recovering spirit she obliged the geese by letting them into the outhouse they had chosen, set about concocting a meal of some sort and waited for Giles to arrive. It was beginning to get dark and she couldn't help wondering why he was so late, when at last she heard the rumbling of Trafalgar and its trailer on the bumpy track.

"Have you got Tacitus?" "I don't know", pointing to the open hatch at the back of the car. We lifted out the kittens, none the worse for their journey, and were beginning to recount our different experiences during the afternoon when we heard an adult miaow, and Tacitus emerged from her nest in the greenery. She stretched herself, reclaimed her family and suggested supper.

## THE GREAT MOVE

The day was already far too long. Nothing but essentials got any attention that evening. While Mary produced some food, human and feline, Giles ran a cable extension out of the kitchen window and plugged it into the freezer, still in the trailer parked outside. We snuggled into sleeping sacks and fell asleep to the reassuring sound of water running into the troughs in the yard, background music that was to accompany our sleep for untold years to come.

## Chapter Four
## UNDER WAY

If fatigue could guarantee a long night's sleep our first day in permanent residence at Bank House would have started nearer noon than dawn. Instead, for the second day running, we were woken early by the engine of an unexpected vehicle. Without even having to rise from our uncomfortable camp beds we saw a small car had driven up and come to a halt in the middle of the farmyard. Out clambered two figures, instantly recognisable as two of Giles's 1969 'A' level history pupils incongruously far from their homes in Hong Kong and Hampstead Garden Suburb. "We've come to help you move in!" they shouted up cheerfully as we stuck our tousled heads out of the bedroom window.

It may have been noticed that in the detailed account of all our preparations we seemed to do little to provide for what was possibly our greatest need of all: recruiting the young people for whom the enterprise was devised and on whose muscle and idealism we were to depend. This omission was partly deliberate, certainly conscious. We had a natural disinclination to advertising and felt it was particularly inappropriate for the time being. How could we know exactly what we were offering or asking of young people until we were up and running? Furthermore, since we had no means of gauging at all accurately how big a job we were taking on, we had recognised it might prove necessary for us to employ someone on full pay for a year or two while we got the farm back on its feet, someone of local farming experience, especially with machinery. We had actually set aside funds for that purpose but we kept on avoiding making any such appointment because we suspected it would not be compatible with the rest of our plans. Would anyone worth paying work seriously on our unconventional lines with young volunteers who might have been taken on for their own needs rather than for their ability? Conversely, was it sensible to expect young volunteers to work hard alongside a mature paid worker while accepting their own lack of comfort and remuneration? We never resolved this quandary but pursued our way in blind faith, trusting that if we lived and farmed as seemed right for us, others would find it attractive too and want to share the experience.

Thus it was that we arrived at Bank House to embark on the restoration of the farm without any definite promise of long term assistance, though we had put out a few feelers. In retrospect we are surprised that we were not more worried by this uncertainty but then security played little part in our

## UNDER WAY

story. Instead our optimism was confirmed by the fortuitous appearance on our very first morning of these two ex-pupils just when they were most needed. While we got dressed they set off down the stony track with a pair of steps, a hand saw and what we called "the mighty cutter". In turns one steadied the ladder while the other stood precariously on the stop step stretching to cut back the overhanging branches that were not used to letting high vans go by. Luckily they managed to finish clearing a way for the removal van from York just before it came rumbling up the drive.

There wasn't much of the day left for any of us by the time we had got all the essential furniture into the house in some sort of order and all the rest under cover in a cow byre which fortunately had not been housing cows! Next morning the couple departed early as cheerfully as they had come, having plugged the only gap there might have been in the sequence of friends and relations living and working with us. As it turned out we were to wait until December 23rd before we spent a whole day without any resident help. Meanwhile we came face to face with the challenge we had set ourselves.

Where to begin? As so often in farming the question was determined for us. Ours not to reason why or suggest alternatives but to accept each urgent priority in turn and get stuck in. Although we were entering the second half of August the first answer was "hay making" or rather "hay timing". Our plans for the 1973 hay season had been laid in our winter visit in January based on local advice which assumed that since we were not due to move in until August 13th we should not be here in time to make our own hay. So we had arranged for a neighbouring contractor to do it for us and expected to find it all finished when we arrived. We therefore left the business of buying our own mowers, tedders, balers etc until the comparative leisure of the farm sales in the autumn – which we were soon to think of as the "back end".

As luck would have it that summer had been unusually wet until the heat-wave which accompanied our arrival, a weather pattern doubly fortunate for us. In the first place this late burst of very hot days produced some of the best hay of the season. Besides this we were unexpectedly present to watch while the contractor dealt with the second half of our hay crop. We were able to help to some extent especially when it came to stacking bales on trailers and in the barn, so we picked up a great deal of general know-how and local experience: when to cease baling as the evening dew begins to fall; how to construct a well-knit stack of bales; how to rope on a trailer load of hay so it survives the bumpy ride up the steep stony track from the lower fields to the Dutch barn, using a secure knot that could yet be loosened with a single tug in the right place. We watched the skill and resourcefulness of a long tradition of dales farming and saw various ways of rescuing tractors and balers that had got bogged down, so soft were the fields under their thin crust of sun-dried surface.

Four long, hot days of shared exertion brought haytime 1973 to a close with eight of us reclining on top of the completed stack, emptying the beer from a large ewer we had taken to be filled at the nearest pub. That Saturday night the elements conspired to emphasise and encourage our sense of achievement at our first job accomplished. It seemed a good augury that the heavens were on our side as broiling hot sun gave way to steady rain through the first afternoon when we no longer had hay lying out to spoil. It was our first pause and marked with almost biblical precision the end of our first week. "Six days shalt thou labour."

Though haymaking was the major achievement of our first week and took priority over other demands on our attention whenever our assistance was needed, it was by no means our only work. Even in those first days we were introduced to most of the main problems that we should have to tackle in the year ahead. Each day threw a series of different challenges at us that could not be ignored, though usually only immediate, short-term remedies were possible. No sooner had we begun to attend to one matter than another intruded with greater urgency. While hand raking a field we would notice that the geese had found their way into one of our two fields of ripening barley. But while returning the errant geese to the yard we saw that a few of our sheep had managed to cross the cattle grid and were sampling our neighbour's grass. The cattle grid needed weeding at once so the gaps between the bars were revealed and would again deter the trespassers. However before that weeding was complete word would come that all hands were needed to load newly baled hay onto the trailers. But some helpers were occupied well out of earshot. One was stalking the heifers high up the bank trying to identify individual beasts by reading their ear numbers with the aid of binoculars, a task rendered all the more difficult because of the Devon Cattle Breeders' tradition of marking their cattle with numbers tattooed inside their ears, and all the more hazardous because after six weeks on their own our herd was too wild to let us get close enough to read the numbers with the naked eye. Another person would be in the back garden beginning to repair the ravages of the recent incursion and planting replacements until interrupted by the delivery in the yard of a big order of fencing materials. The driver – and his two dogs come for the ride – needed help unloading thirty rolls of barbed wire and fifty rolls of pig fencing so the

## UNDER WAY

planting was postponed only for the storing of the fencing materials to be stopped in its turn until the day's hay had been stacked under cover.

One evening the contractor's son came to announce that baling was due to start the next day in the field called Ludla provided they could get the baler into it. The culvert leading into it was blocked underneath and a foot or more of water was streaming over it. Working in rushing water we had to pull debris of twigs, rotten leaves, small stones and gravel from either end of the culvert and then rebuild the collapsed stonework of the culvert itself and deepen the streambed downstream of the culvert, hoping the track would dry out enough overnight and solidify to the point it could bear the weight of tractor, baler and wagons of hay. This urgent job took us away from another of long-term importance on which some of us sweated for part of almost every day that week and well after.

Black Carr was the poor field where we wanted to build our complex of barns and cattle yards. Though our permission to build there was based on our builder's proud assertion that he could build anywhere, it remained for us to drain this boggy field enough to justify his boast and our optimism. The previous winter, and again at Easter, we had seen water pouring steadily out of a land drain some fifty feet up the bank above the projected site. It was this supply that had led the ministry drainage advisor to be confident that we should have ample drinking water for any cattle yards we put up in that field. But that constant stream of water was pouring night and day into the very field we had to dry up, so we had to dig an open ditch to divert it round the site into the half mile length of gutter which was the main artery of the farm's drainage network and which we had got contractors to reconstruct that Spring, an earthwork we christened "Offa's Dyke".

Every day we took spades, buckets and crowbar to deepen our diversionary ditch, soon known familiarly as "Black Carr Corner". Every morning we found the upper bank of it had performed a series of mini landslides overnight and filled up our channel with what we called "oojah" for want of a better term, so the water was heading once again for the building site. We felt ourselves engaged in a primeval struggle with the forces of nature. To make matters worse the combination of hot sun and watery sludge proved ideal breeding ground for midges and mosquitoes. We were severely punished if we failed to put in our day's stint of mud-bath excavation before the onset of evening brought the pests out in full strength. It was an unforgettable taste of Flanders warfare. We knew we were engaged in the early stages of a lengthy battle, however, by the end of that week there were encouraging signs that it was a battle that could be won.

Alas Black Carr Corner was not the only scene of relapses. On our first Sunday, being appropriately our first day of rest, some of our heifers decided to pay a return visit to the kitchen garden and ate or otherwise destroyed the greens that we had planted on the Wednesday, our first attempt to make good our lost winter's vegetable supply: like snakes and ladders, back to square one. It was as though the heifers knew that we were at last about to reconstruct the garden defences having finished haymaking the evening before, and as if they had to make their own kind of hay while the sun shone.

There was much elation and satisfaction in those early days but there was a fair measure of grief and disappointment too. Learning the hard way we had to take the rough with the smooth. We found one of our geese dead that same Sunday. Poor Dora ! We had not noticed her losing weight, having assumed that with so many legal acres to graze, quite apart from their illegal excursions into corn fields, the geese could forage satisfactorily for themselves. Too late we realised that most of the grass on the farm was too coarse and lacking fertility to nourish our geese adequately. Possibly too, lack of familiarity of the farm's geography may have stranded them without water for hours at a time in the broiling sun.

The certainty and detail with which the events of these early days can be related over thirty years later is due to the daily diary which Giles kept all the eighteen years we lived at Bank House. It was begun as a necessary practical aid to our work. Every day we were receiving a wealth of information and advice which it was essential we should not forget, dates for the farming calendar, when to plant or harvest crops, which contractors, vets, or tradesmen would best serve our purposes, where to buy such and such, where to get this or that repaired, how long it was since we began grazing a particular field, and so on. Almost at once the scope of the diary widened to include social matters and personal reflections, and since there was little time to write letters it became a habit to post pages of the diary instead as a way of keeping in close touch with family and friends, and so it became worthwhile to add items that lent colour to the picture it conveyed whether or not they held any agricultural significance. The following extracts all come from the pages of that first week's diary:-

"*Mary saw an owl fly out of the Dutch barn*"

"*pancakes for supper*"

"*low flying aircraft hired by our neighbours, sprayed bracken – acrobatically*"

## UNDER WAY

*"Aga too cold for roast beef – riddled – too hot – burnt Yorkshire pudding"*

*"Herodotus and Cleo (kittens) woke us standing on our faces, tapping light switch against the wall"*

*"postman says honey now ten shillings a pound and has a swarm for us".*

Such entries tell the story of our settling in and mark many aspects of our early progress though sometimes it is not easy to decipher the scribble trailing off into wandering lines as Giles often fell asleep in mid sentence at midnight. It is no surprise therefore to find for Wednesday August 15th "G began writing letter to CB", and the next evening, "still not finished letter to CB". It must have been important to be bothering with writing when so much needed doing urgently outside. It was in fact our attempt to draw up the terms on which we were inviting this teenager to come and join us. It was sent to our first candidate and must have proved satisfactory for four days later the diary records a phone call from CB's father saying the lad would come for a trial period the following week as our first long-term helper.

Of course we too were on trial. Not only had we to learn how to apply our organic principles while farming on a shoe string, we also had to stretch that shoe string to create an acceptable pattern of life for our helpers, avoiding anything smacking of the culture of school or parental authority. Naturally we failed at times. We had our noses so close to the grindstone we cannot always have been easy to live with. It was much easier while our early helpers were mostly short term holiday makers and often our own relatives as well. Something of the holiday spirit helped to keep up our morale however hard we worked and the sense of adventure buoyed everyone up, which may explain why Giles's father absent-mindedly addressed a letter to "Bank Holiday Farm" after his first visit. For that first week or two we ourselves could not help feeling that we were having a wonderful, active holiday and that we might wake up next morning to find it was all over and that we were due back at school with essays to mark and pupils calling us "Sir". Very soon however the urgency and responsibility of the needs of the farm became all engrossing: it was memories of school life that faded into ever more distant unreality. Even our night time dreams changed their setting from Abbotsholme to Bank House, and their themes from getting pupils through exams to getting sheep through gates. In the world of dreams these were not always so very different.

Here the account of our first week at Bank House comes to an end. To put progress so far in perspective with the mountainous task ahead we have only to remember the ten years we estimated it might take for the farm to recover thoroughly from the shameful years of neglect. In other words, of the possible five hundred and twenty weeks of work we had now completed just one.

Many of the strands that will make up our story have now been introduced but even more remain to come. Indeed they are branching and multiplying at an increasing rate. As with a juggler adding ball after ball to those already in circulation it will soon be impossible to keep all of them in collective focus. From now on we shall have to concentrate on one strand at a time but before we do so we shall round off this predominantly chronological part of our record by quoting the round robin of greetings we sent out to all our friends and relations for Christmas that year, 1973. A very limited de luxe edition was embellished with free-hand sketches of the farm perched on its bank beneath its moorland skyline.

### GREETINGS FROM GLAISDALE!

*This is such a beautiful valley to live and work in, we long to share it, to show you around and write at length, but with only forty-eight hours a day (between us) we may be forgiven this duplication of correspondence – just this once.*

*If you walked in today and saw the mess that the house is in, you might well wonder what we've been doing since August 13$^{th}$. Nothing decorated, no door handles, small holes in the unplastered walls where a damp course was injected, and not a curtain in sight. If you peered into the cow byres across the yard, you would see boxes galore, stacked in piles amid beehives, trunks, furniture and rolls of carpets; we have only half unpacked.*

*Yes, in terms of particular items progress is slow. The new water pipe from the reservoir in the rocks up the bank (promised for hot August) was laid last week in the snow. The new roof and dormer windows should have turned our attics into rooms with a view in September, but the tiles are stacked in the yard, the scaffolding is up and any day now we may see the sky above our bed. The cattle grid has been propped against the wall down by the road since October; with luck you may not have to get out to open the gate by the time you come.*

*Some days we look at the holes in the track, the piles of stones that should be walls, the gateways without gates, the hedges running riot, the ditches choked and overflowing, and we wonder how and when and if. But much more often we are impressed by the vast amount that has been done. Thanks to all sorts of help.*

# UNDER WAY

Out of doors, eighteen red Devon heifers and a September calf took the recent frost and snow in their stride. Winsome now leads a flock of 59 ewes with a new paramour from Kent, Ethelbert. Monsieur Mollet's gaggle of sixteen is being diminished in the interests of Christmas, but the farmyard population is augmented by Autolycus,, our white bantam cock, and an elegant ensemble of hens; while Kim, our six-year old sheep dog sleeps under the chicken loft steps in his newly made kennel, on sentry duty when he is not teaching us how to manage the sheep. Tacitus prefers to stay indoors when not hunting mice to train Tiberius as a real fighter (Gracchus not the Emperor).

What else have we done? We went to countless sales and hauled tractor loads of "dead stock" home over the moors; a muck spreader, corn drill, plough, two mowers, harrows, railway sleepers and so on. We have fenced, and fenced, and fenced, so now we can actually choose which fields to keep stock in. The lower track is almost dry, thanks to ditches and cuts and stones. There's a beautiful thatched straw stack to witness our efforts; likewise a huge pile of logs and eleven boxes of apples, a workshop with a bench and vice installed, stacks of timber, and naturally, a couple of waxing compost heaps for farm and garden.

Our first four months here leave us with a wealth of memories and impressions; the excitement and reward of learning, fresh problems and new skills; the heat-wave of our late haymaking, gathering hazel nuts in the mist of an October morning and blackberries in the sun of the same afternoon; digging in the mud for vanished drains; the fierce winds off the moor rising with the fall of evening and yet picking dahlias in mid-November; searching the wooded ravine beyond our boundary for four errant heifers in five inches of snow; the humour and excitement of farm sales.

And perhaps, above all, the ever-changing beauty around us; the water running in the troughs in the yard; the fresh air; the colours of bracken and tree, grass and stone. Never idle, sometimes alone but never lonely. We have been lucky to have C. B. (16) living and working with us since the end of August.

Our visitors' book has fifty-one names in it, and we hope yours will be added before long — but don't forget to bring your gumboots!

Mary and Giles with helpers on the chicken loft steps

## Chapter Five
## "FENCING, FENCING, FENCING"

In this district the pioneer process of carving farmland out of the wilderness of woodland, bog and moor was principally achieved by building stone walls. They staked their claim, provided shelter and cleared the fields of loose, and some not at all loose, stone. The labour involved in their creation was prodigious and it is not at all surprising that people in later ages resorted to easier alternatives. Today walls survive chiefly on higher ground while hedges have largely taken their place lower down, very often on the same lines, so that much hedge planting is bedevilled by the foundations of vanished walls.

Hedges growing year after year need more maintenance than well-built walls and the practice that produces the best stock-proof hedges is that of laying or layering, which is necessary only every now and then. It is a highly skilled activity like dry stone walling. Both stone walling and hedge laying require years of practice so that it is hardly surprising that very sadly we decided to postpone work on these activities for some years, apart from a few experiments here and there.

In this as in almost every sphere of farming, technology has given us easier, quicker and cheaper alternatives. Unfortunately they are often more ugly too. But the availability of factory-produced wire fencing and the uncivilised invention of barbed wire made it possible for us with our unskilled volunteer hands to reclaim a derelict farm quickly and cheaply. So it was that, a little shamefacedly, we set about our huge urgent task of fencing.

The man from whom we bought Bank House had been a butcher. It seems that he shared the belief common to man and beast that "the grass is always greener on the other side." In his case this led him to think that the farmers breeding the animals he bought had the easier life, simply watching their stock feed themselves. That was before he bought the farm! Then for nine years he got away with neglecting the maintenance of most of his stone walls, hedges and fences because he had no cattle, sheep or horses of his own to demonstrate that where posts or rails rot, where hedges die back or where stones fall off walls, single adventurous inmates can rapidly initiate a tide of illegal migration, widening the breaches disastrously, as we were soon to witness to our cost.

# "FENCING, FENCING, FENCING"

Restoring cattle and sheep to Bank House was rather like putting fish into a net with too large a mesh. Every outlet needed closing at once. For much of our first year we were repeatedly having to halt work at one place to dash off and plug a new gap somewhere else before it got a lot worse. We had to divide our efforts between making our outer boundaries secure – which had the double advantage of keeping our stock in as well as our neighbours' stock out – and restoring internal field divisions so we could control which animals grazed which fields or which could be kept apart – one tup from another, a bull from his daughters. The urgent need to fence off a bog, ravine or a hay stack often settled the matter temporarily.

Such was the poor state of these internal defences our first winter that the herd had a third of the farm to roam at will, including the twelve acres of the Nutwood. They could not use the next thirty acres for there was little to stop them wandering off along the edge of the ravine and through the West Arncliffe Woods before reaching the open road to Whitby. We recognised that the task of making the edge of the ravine safely stock-proof was too big a job for our first year or two so we decided to separate off the furthest twelve acres with a completely new fence running downhill at right angles to the beck. This necessitated erecting a stretch of fencing through some hazel wood so steep it was barely possible to keep one's feet while at work. One of us would hold a post in position with one hand while clinging to a hazel stem with the other. His partner, similarly hooked onto another hazel clump, would use a crowbar to make a hole in the right place and then bang the post in, operating one-handedly in each case. It is remarkable that the resulting amateur-looking fence still serves its purpose over thirty years later.

Elsewhere other problems made fencing difficult in different ways. Generally Bank House is not an easy place in which to drive posts into the ground. Hidden boulders or soft bog can be equally baffling; so can the roots or hindering branches of trees. Nor was it a simple matter getting the tools, posts and rolls of netting and barbed wire where they were wanted. When fencing the top of the Nutwood that first winter we had to be our own pack horses up the steep bank to about seven hundred feet. We used to push a post through the core of a heavy roll of wire netting and set off uphill with either end on a sherpa's shoulder. Alas the roll kept on sliding down, threatening to decapitate the lower porter. Reels of barbed wire, thick straining posts and struts, were hardly less awkward to manhandle so that together with the standard five-foot posts and all necessary tools, assembling all the equipment was the greatest drain on our energy on many a day's fencing up the Bank  before we could begin putting up the fence itself. Even then there was still much to do preparing the site: rushes, brambles

## "FENCING, FENCING, FENCING"

and branches, even whole trees, had to be cleared away and the ground levelled and drained if the posts were to be firm and the wire netting close enough to the ground to prevent lambs and small piglets from crawling underneath. It was not surprising therefore that it would take three of us working most of a short winter's day to average no more than six or seven yards of new fence when enclosing the Nutwood.

The most frustrating fencing however, because the least effective and the least likely to endure, was along the banks of the meandering beck, lined all the way with trees large and small. Quite apart from the need to resist the temptation to hammer nails and staples into the trees themselves to the possible future ruin of axes and chainsaws, trees make poor fence posts. They move in the wind, even blow down in gales, tearing away chunks of a bank in the process, and they drop heavy branches on the fencing below. They also provide rafts of sodden logs and other débris that drift on winter floods with the force of battering rams, sweeping aside the flimsy structures laboriously constructed around the drinking places or crossing the beck as necessary extensions of field divisions. What one might call "inland" fences may enjoy a life-expectancy of some twenty years or more. Their inferior beckside cousins are short-lived and constantly need patching up.

This may explain why the diary of our first three months records incursions by a neighbour's cattle from across the beck on seventeen separate occasions despite frequent repair missions by our neighbour as well as ourselves. The reason for this traffic being largely one-way that first season was that most of the time we grazed those fields with sheep and kept our herd well away on higher ground knowing our inadequate beckside fences would not deter them from making return excursions of their own.

Of course progress was greatly slowed down whenever a gateway was involved. For each gatepost a hole had to be dug 3 ft 6 in deep. There are many ways of extracting stony earth loosened by a crowbar or a spit at the bottom of a narrow hole, but none of them is quick or easy. When the post is set in its hole it has to have stone and earth packed hard round it all the way down to make the post rigid. Hanging a gate on its hinge post is a fine art which is why in January for our first new gate we chose a remote spot where few passers-by would notice our incompetence. Fortunately an instructive neighbour came to our rescue and it turned out very respectable. So we learned how to make a gate rise and fall when it swings as the ground level requires. It is a great bonus if a gate will gently swing shut of its own accord.

# Fencing an awkward spot

## "FENCING, FENCING, FENCING"

The long, six-foot-deep, open gutter we named "Offa's Dyke" posed a protracted priority fencing job as soon as we moved in. A condition of the grant we were claiming for having had contractors dig out this main artery of the farm's drainage system was that it be "fenced" on each side. This condition was deemed to have been met by the contractor fixing two strands of barbed wire on widely spaced thin posts. This might have deterred well fed cattle but tempted grown sheep to step between the wires while lambs hardly noticed as they strayed beneath. This left us with over a mile and a half of inadequate fencing to upgrade as soon as possible. We doubled the number of posts and added sheep netting in place of the lower barbed wire which was moved up above it. This was comparatively simple work since half the necessary posts were already in position as was one strand of barbed wire.

Among various difficulties we encountered in our fencing operations the ones we least expected were a landslide and incurable deep bog. At one point in its long course Offa's Dyke ran into more than usually treacherous boggy ground. Despite being warned, the subcontractor on the job drove his JCB digger into it. Our neighbouring farmer's wife took a photo with only the cab roof visible above ground level. A huge caterpillar –tracked vehicle had to be summoned to extricate it before we fenced the area off and planted it with trees. To this day it is referred to as 'JCB's grave'.

Another surprise awaited us at the very top of our sloping fields where derelict fencing and walling separated them from the estate woods above. We had been in residence barely a couple of months when a narrow landslip began to develop, slowly bringing a thick tongue of wet clay and rock across the boundary carrying whole vertical trees with it, some even thirty feet in height. This posed us peculiar fencing problems for many years as each time we placed some new fence along what we judged to be our legal territorial boundary the land continued its inexorable slide, first bending and straining, then sweeping away our most recent handiwork. Of these two unexpected problems, JCB's grave was the easier to deal with though the more dangerous. The landslip proved the more awkward though the less threatening for cattle and sheep which always were wary of the quivering ground around the landslip and never attempted to exploit the gaps periodically appearing in the fencing. Nor was the dark wood above particularly appetising. Luckily for us the grass is not always greener on the other side after all.

It might have been expected that this vast amount of fencing and the practical problems it posed would have had a daunting effect on those engaged in it. On the contrary, it was an asset. It provided a permanent source of constructive activity with a wide variety of subsidiary jobs all of them clearly necessary and rewarding. They could be carried out in gangs or in ones and twos, so we could always occupy whatever mixture of visitors and volunteers happened to be available. It was comparatively easy for us to match the skills and stamina required to the work selected. Furthermore, it could be continued from day to day so that people began to identify themselves with a particular project. Some would return to it on subsequent visits with pleasurable pride. On the other hand work on a fence could be shelved, sometimes for weeks, to allow for changes of scene, company or weather.

At times half a dozen or more people could be seen variously engaged along the line of a single fence. At one end someone would be digging in a big straining post or making neat joints for its supporting struts, while someone else would be laying out posts, one to every three yards, which others were driving into the ground, working in pairs if the line was to be straight and the posts upright. Others again might be levelling awkward undulations or flattening out 50-metre rolls of wire netting and joining them together, one of the hardest jobs to do well. It is surprising how far a fence is made kinky by a poor connection. Every one liked letting fly with hammers and staples when it came to fixing the netting and barbed wire on to the posts, as suddenly the new fence stood erect and serviceable. The apparently simple act of banging in staples provoked a competition to see who could drive staples fully home with the fewest number of blows. But staples have a will of their own and competition often ended in scratching around in long grass searching for elusive staples so that cows did not wrap their rasping tongues round them instead.

One unique error remained in full view for decades. Though close alongside a public bridleway it is unlikely that any passer-by ever spotted it. A helper, noticing that with standard sheep-netting the rectangular mesh was graded in size, fixed up a whole roll of it upside down, with the biggest gaps next to the ground and the smallest at the top. He hadn't thought it out that smaller animals need smaller meshes to contain them, whereas any creature tall enough to reach the top of a three- foot fence would still be excluded by the bigger mesh. Considering where it was we decided it didn't matter and left it intact. The only difference it would make was that grateful rabbits would find it easier to race through the fence without slowing down.

In a way this mistake can be seen as symbolic of our Bank House venture as a whole. Where else in the dale, nay, in the North York Moors, would such a mistake have been made and deliberately left unrectified as a monument of respect to the efforts of a young volunteer?

## Chapter Six

## WINTERING CATTLE IN THE OPEN

On coming to Glaisdale we found it general practice to winter cattle under cover. This is as much for the protection of the land and the convenience of the people involved as for the well being of the animals themselves. The timing of bringing them in varies each year and from farm to farm according to factors unique to each case, but the second and third week in November will generally see a great many taken in. With good food and a degree of rough shelter during spells of unusually severe weather most beasts can stay out considerably later without coming to much harm, indeed even in this northern district a few herds remain out all winter. It had been a point in favour of our choosing the Devon breed of cattle from the Exmoor district that they were notably hardy in this respect.

So it was that as our first winter approached we were not unduly apprehensive about our lack of cover. We were well aware it would be tough going without it and were determined that we should not allow other pressing demands on our time and energy make us forget to push ahead with getting our cattle yards built as soon as possible. Meanwhile our heifers would winter out.

Quantity of food did not seem a particular problem, even with our slender bank balance. Our own recent hay harvest was roughly trebled by the hay of unknown vintage left by our predecessor, and the Dutch barn was stacked to the roof. Furthermore, a great deal of the farm was covered with a mat of course grass and weeds, the accumulation of years of neglect. It was the essential function of our cows and sheep to graze this down and in so doing to launch the process of improving the fertility of the soil and the quality of the sward.

Our chief difficulties arose from the state of our walls and hedges and from the fact that our heifers were strangers to us, to each other and to the farm. In later years we had only to open a gate and summon the herd from a distance for it to gather behind a leader and break into an accelerating tide, eager for fresh pasture or ration of hay. Alas we enjoyed no such labour-saving habits of cooperation that first winter. Having been assembled from four different herds, our heifers remained for many months in distinct groups one of which quickly established a dominating authority from which the others escaped as much as possible. This made our daily inspection time-consuming and tiring, making us search for four herds instead of one, scattered widely, some in the twelve-acre nutwood, some in the banks on either side and others yet again possibly right out of sight on our highest ground where there was no guarantee of finding them even if we clambered up the intervening two or three hundred feet.

That was merely a matter of perseverance. Much more difficult and worrying were the occasions when we had to bring them indoors where they could be handled and identified with absolute certainty, as for the Ministry Inspector checking our claim for a Hill Subsidy. Most daunting of all was the prospect of the compulsory annual TB & Brucellosis tests in November. We lost quite a lot of sleep beforehand wondering how we could guarantee to collect all our heifers into stables and loose boxes in time for the vet to carry out the tests, for which every animal in turn had to be held fast for the vet to take a blood sample and give an injection. Every well-equipped farm had its own cattle crush for the purpose or shared the use of one with neighbours who had clubbed together to spread the cost. Our preparation included getting permission to borrow a crush from such a group up the dale only one of whom could remember where it was last used, and having tracked it down we mounted it on our own tractor and brought it home the day before to set it down in position adjacent to a barn or loose box door, a manoeuvre that took time for novices because of the great weight of the crush and the uneven ground in our yard.

We were ready. Not so the heifers, three of whom could not be found. Searches all round the farm the night before, and again early in the morning of the tests, were in vain. We even took the car and a pair of binoculars across the dale in case the missing trio were visible from there. At the eleventh hour we found them in the Estate woods beyond our boundary and managed to drive them back to our yard just in time. It proved as difficult as expected to get each heifer into the crush. Two broke away and had to be rounded up for a second attempt. Eventually we coped well enough except with the misnamed "Helpful" one of whose horns stuck straight out at a threatening angle and she knew how to use it. She knocked the vet to the ground and tried to gore him like a bull. Lying flat on his back he bravely grabbed a horn in each hand and, forcing her head away, managed to scramble to safety.

That was the end of round one, ending in relief rather than victory. Three days later the vet was due again to check the reaction of each animal to its injection, so once again we were supposed to get each heifer into the cattle crush. Having kept them all close by during the night we were able to get them into the stable in good time in the morning, but when we placed the crush across the doorway bovine memory stirred. Cattle crush. Jabs and

# WINTERING CATTLE IN THE OPEN

needles! Mind out, girls! Nothing we did could induce the bigger beasts to enter the crush a second time so in the end the vet had to manage with them on the loose. We were relieved again to have got through, relieved too that we did not yet possess a bull.

Wintering cattle outside had other distinct drawbacks of which the greatest were the use of the AI service and feeding hay on the ground. The agreements we made when purchasing our heifers stipulated that they should run with the bull until they came to us so that with reasonable luck we should have had eighteen heifers calving in February or soon after. Whether we just had unreasonably bad luck or whether there was an element of mismanagement we shall never know, perhaps both, but to our growing dismay first one heifer started bulling, then another and another, proving that they were not in calf. Each time we saw this happening we had to get the lady concerned up to one of the loose boxes in the yard and ring for the AI man to come and serve her. Since there were only two or three days while she remained "in season" there was no time to lose or we should have to wait another month; another month of feeding without a calf on the way and ultimately a younger, smaller calf for the autumn sales. Because most of our heifers were still too scared to be separated completely from their companions and driven to the yard alone we usually had to bring several others too and we ended up taking up two or three people away from other valuable work. In all we had to call on the AI service seven times that season. In years to come when the stock would be housed in fold yards under constant supervision we couldn't fail to notice any bulling and the herd would be conveniently close at hand on the fewer occasions we needed the AI man; "fewer" because by then we should have our own bull.

Haying cattle out of doors on the ground was to be another job taking up a lot of our time that winter. As our sheep and heifers ate off the best of the cover of old grass, and especially when there was frost and snow, we had increasingly to put out hay on the ground wherever the animals were grazing. Whenever we finished making one of our lower fields stock-proof (comparatively!) we were able to take the cattle down to graze it, and as the grass diminished we would supplement it with hay. Regular use of the tractor to do this soon churned up sections of the track creating another job to do, shovelling soft mud off the track and using it to fill up awkward hollows in what used to be level hayfields. The wetter the weather the more difficult haying became, dodging the showers and trying to find shelter to put the hay along the lee-side of a hedge or a wall, hoping it would be eaten before it got too wet. There were times when the only shelter available was provided by the bodies of a tight bunch of heifers. Some wastage was inevitable but until we got into the new year it worked well enough and enabled us to move the haying about and avoid damaging the ground too much in any one place.

Welcome side effects of haying the heifers were that we began to get to know them individually and that they began to get much tamer. Unfortunately this did not curb their taste for adventure which was demonstrated only too clearly as November ended with snow and ice. A letter to Giles' parents dated 2$^{nd}$ December recaptures some of the drama:

> "It began with terrific gales that nearly demolished a haystack: the black plastic cover was flapping violently and lifting the heavy bricks and logs we had tied on to weigh it down so they danced about like dangles on a baby's pram. A combination of tying the sides with nylon cord, hanging heavier logs on and tightening up the whole net that's on top of the plastic sheet all helped, though it looks a bit like a patient emerging from a casualty ward – crutches and bandages.
>
> The wind blew where it listeth, alright! The stack listeth too! Four of our heifers took themselves off to the shelter and possibly food of the wood beyond our boundary and don't want to return. We found tracks leading off so we had to go searching and in doing so explored the ravine and slopes all the way to Glaisdale station. We lost them completely for four days; made two big sweeping searches. It was a fascinating experience following footprints and prodding cowpats to see if they had been frozen all night or dropped recently.
>
> We also made contacts socially. Mary called on two farms on the other sides of the wood, people we don't normally meet and one family took great care and interest. The chap everyone calls "Tim" went walking about with binoculars and rang up to find if we'd got them back. Well we haven't, not to date anyway. But we found them, very happy thank you! they prefer their part of the wood to ours while this arctic spell lasts and they can dodge around the bushes and tree trunks, gallop over snow-covered bracken and up the slopes so much more nimbly and tirelessly than we, that our efforts to drive them through the open gate failed entirely. We've taken down a fence in a much more attractive, enticing corner to drive them through the next time we try but it's too exhausting to do it every day – you have to climb four hundred feet just to try. And they seem fit where they are."

As if that wasn't enough excitement, two days after that letter was written a huge vehicle loaded with seven tons of gravel struggled into our

## WINTERING CATTLE IN THE OPEN

yard by mistake and got bogged down. Apparently the driver thought we looked as if we could well be called "Browside Farm" where in fact it ought to have been delivered. Our tractor couldn't shift it and it took two more and a borrowed hawser to tug it free, leaving holes that were unlikely to mend until the following summer. That evening however, as if to prove that fortune can be even-handed, the four errant heifers wandered down the bank into our yard to appreciate some hay. Next day we reunited them with the rest of the herd in a small field with completed fencing.

## Chapter Seven
## FARM SALES

The work of the hill farmer, like that of the proverbial housewife, is never done. Ignoring every other claim on our attention we could have filled every day of our first year doing little besides feeding animals, moving them about and of course fencing. Fortunately there can be no danger of monotony on a mixed farm, the inexhaustible variety of necessary activity sees to that, so if ever interest or energy is in danger of flagging there is little difficulty in justifying a switch to another occupation using different skills, and muscles and enhanced by a change of scene. Nothing proved more effective in this way than going to a sale.

In the normal course of events when a farmer retired at the end of his career he would arrange for an auctioneer to sell off all his livestock and equipment – everything that is which could be moved. Unless he were a tenant the sale of the farm itself, land, house and other buildings – would have taken place previously and quite separately so the new owner had the chance to bid for the animals and machinery which presumably were suited to the needs of that particular farm and were already on the spot, thus saving a great deal of trouble and expensive transport. Alas we had no such advantage when we bought Bank House where our predecessor never held a public sale at all. This was because he possessed extraordinarily little in the way of animals, tools or farm machinery having relied on contractors for such haymaking or arable work as took place. Instead he disposed of his few movable assets piecemeal, even selling off some of the gates after we had bought the farm! When we took over the cupboard was bare and we had to re-equip the farm virtually from scratch, buying everything from tractors and trailers down to hayracks and pitchforks. By far our greatest expense was to be acquiring good livestock, and building our complex of barns and covered fold yards, which made it essential that for everything else we had to manage as cheaply as possible. As a result all our farm machinery was at least second hand. In all the years running our farm we acquired most of our hand-tools and lesser ironmongery – nuts, bolts, hinges and the like –as bargains picked up in strangely assorted boxes at dispersal sales.

It might have been expected therefore that we would have to wait several years before finding farm sales that offered all the equipment we needed, and more years again for us to succeed in buying it, leaving us an awkward interim of borrowing or making do with unsuitable improvisation. In the event we were extremely fortunate that there were more sales than

# FARM SALES

usual that autumn being held within easy distance and advertising just the things we wanted. This good fortune was due to the state of English farming as a whole at that time. The process of small farms closing down and being swallowed up by larger neighbours was well under way; we had examples in our own small dale, indeed if Bank House had not been in such a dreadful condition it would have gone the same way. It was also a period of generous agricultural subsidies which encouraged a lot of farmers to expand their undertakings and in most cases this involved greater mechanisation, larger machines for larger fields, and less manual labour, a tendency exaggerated by a widespread change in farming practice as silage increasingly took the place of hay. Any enterprising couple fresh from agricultural college setting up at that moment would have equipped themselves very differently from what we planned to do. We bucked the trend quite deliberately and made haymaking the basis of our farm strategy. As a result we were glad to find sale after sale offering a wide variety of old hay-making machinery which was fast becoming out-dated and redundant. Most of this was designed for work with tractors as working horses were already being consigned to history, but farm sales have a habit of unearthing long-unused items which farmers have hung onto from a mixture of inertia and nostalgia. Though we didn't yet have a horse we were already on the lookout for horse-drawn implements and harness. However when they appeared their price was usually driven beyond the scope of our budget by trendy publicans and antique dealers instead of leaving Mary to contest it at a reasonable level with the scrap iron merchants.

We soon learnt that if one relies on farm sales as a source of supply there needed to be plenty of them. Luckily that year there were. In our first three months at Bank House we went to no fewer than thirteen local sales, usually taking any helpers or other visitors staying with us. Naturally no-one wanted to be left behind for sales were great social occasions imbued with the holiday spirit of a community of farmers released from the daily round of their isolated existence, a mood we readily shared as a break from our continual fencing. The sharp wit of auctioneers, particularly when provoked by answering wisecracks from the crowd, provided entertainment to rival the music halls.

Mary had had experience of auctions long ago in Sussex and more recently near Abbotsholme but for Giles and our young folk sales provided important education. It was a chance to see a huge variety of strange farm tools and machinery in the company of people who had used them, people only too willing to explain the merits and drawbacks of each item, often illustrated by anecdotes that put their verdict beyond doubt! We were told what we ought to look for instead and where else to get it. We made all sorts of useful contacts which we could follow up when appropriate and began to know whom to trust for impartial advice. Gradually we built up a scale of values so we could judge better when to stop bidding. Even so it could be nerve-wracking at times. We also learned to recognise the compassionate aspect of sales never more evident than when they had been occasioned by financial distress, illness or even death. Most people present knew that "there but for the grace of God go I" and dug deeper into their pockets than was their normal thrifty practice, or bought what they knew they didn't really need.

We found sales very time-consuming affairs especially if we were after something important to us and wanted it badly. Then it was worth going the day before when everything was already laid out in a field for inspection and to make sure that the object of interest was really what we needed and in good condition and to learn from the owner how to use it and maintain it.. Then on the day of the sale we would be off in good time to be sure our target wasn't sold before we got there. After buying anything one of us had to join the queue before the auctioneer's clerks to pay for everything we had bought. Since auctioneers deliberately kept all the star attractions to be sold last, such as tractors or even combine harvesters, the bulk of the crowd stayed to the end and then left in a rush and formed a traffic jam. There was usually a lot of delay in getting away with long hold-ups in wet weather with overloaded vehicles spinning in the mud. More often than not by the time we got home, hungry and with animals still to feed, we left unloading our booty until the next day.

Our first season of attending farm sales was a most prolific one. The list of acquisitions was long and extremely diverse. As the months and then the years went by we ceased to find it necessary to go to every sale within range: nevertheless our accounts, verification of so much of our story, bear witness to a continuing line of purchases through our later years of which we no longer remember each occasion. The general description of farm sales given above drew on a sort of composite folk memory in which all sales we ever saw merged into one as a vital piece of northern rural culture. The more sales we took part in the less we recall them individually. It is those early experiences that are unforgettable, when everything was new and exciting, when our needs were urgent, when we had everything to learn and when the survival of our Bank House venture was by no means secure.

Of course one remembers a first occasion. Exactly a month after our historic arrival at the farm we felt ready to go to our first farm sale from

FARM SALES

Bank House. It was to be held at the far end of the North York Moors and the advertisement listed several things we certainly needed so we welcomed the excuse of 'farm business' to explore the further scenery of our new environment for the first time. We set off in "Trafalgar", our estate car, pulling our invaluable little sheep trailer, stopping on the way at Guisborough where we bought a strong dustbin and fortified ourselves with a supply of pasties for a late lunch to be eaten whenever the sale should allow. In spite of drizzle and low cloud it was a wonderful ride with the dramatic northern escarpment of the Cleveland hills silhouetted against the sky above us, and then the climb up onto the moors that looked down on Osmotherly. If we hadn't succeeded in buying anything that day we should still have thought the trip worthwhile, it was such a treat to be exploring the glorious country that had been one of the reasons for our coming to Yorkshire. Our holiday mood was increased by the anticipation of the first visit from Giles' parents since we moved in. For this Giles had to leave for a while to go and meet them off their train at Darlington station and bring them back. It was always an extra pleasure seeing things through the eyes of one's visitors and sharing their reactions. On return we found that Mary had been profitably busy accumulating a pile of lesser bargains. For a mere £3.90 we now possessed a large grind stone, some slashers and gripes, a halter and a bale fork besides useful ropes, chains and augers. But our real prizes were a Massey Ferguson mowing machine listed as a cutter bar, and a two-furrow plough, both of which were to serve us well for many years to come.

Mary had also spontaneously dared to commit us to embarking on another aspect of our plan for mixed farming. She had bid for and bought three bantam hens and a cock. They remained shut in an old shed until we were ready to set off home. By then we had realised that our little trailer couldn't safely carry both the plough and the mower so we arranged with the farmer to come back for the mower another day. The smooth passage of our first sale foundered somewhat at this juncture when Mary went with him to collect the bantams and put them into a sack for our journey home. Three birds were captured but the fourth, an intelligent speckled hen, flew fast and straight at the wooden wall. Open Sesame! Magically two planks of the weather boarding parted to let her through and she winged her way noisily into the distance. A second visit was going to be doubly necessary.

The sale and the scenic journey home with five people, three birds and the travellers' luggage in the car and the heavily-laden trailer bumping along behind gave the parents a splendid introduction to our new life. Next morning for the first time at Bank House we were woken by the crowing of a cock. It took us a moment or two before we realised it was ours.

The return visit to Snilesworth the following week turned out more rewarding than we expected. Loading up the mower was soon accomplished because the farmer had taken it up the steep track to the moor road to make our task easier. Furthermore he had found a spare blade for the mower and the stand on which to sharpen it which we were able to buy cheaply along with some windows that we saw languishing against a wall left unsold at the end of the sale. One of these soon embellished our bathroom and later others came in handy giving more light to the workshop. To cap it all we were presented with the Houdini bantam together with an egg she had laid since her return.

The following week was memorable for a variety of reasons one being simply the spate of four sales in eight days. Two were notable for their lack of anything really worth going for. The diary records "Westerdale a bit disappointing. Main items no good. Bought chain harrow (£18) tarpaulin (£3) and milk strainer (5p)". A similar entry appeared a few days later. "Swainby – pretty ropey – filled car and mini trailer with panes of glass, oddments of netting, posts, guttering and a water butt + two slabs of marble – total cost 90p!" However the operation of the law of swings and roundabouts redressed the balance in our favour as the other two sales between them provided us with four major acquisitions adding to our store of heavy equipment; a muck spreader (£130) a large trailer (£155) – still in frequent use 30 years later – a cultivator known as a "triple K" with rows of S-shaped spring tines (£66) and a versatile hay turner delightfully called a "Vicon Acrobat" (£40) about which a comforting neighbour was to remark that "you can't half make yourself look a clown on that" when he observed our early attempts to use it.

Bidding for something at an auction is the simplest part of the business, be it a cow or a teapot. A mere wink or a stifled nod may be enough to make it yours. Often the real difficulty begins when it comes to getting the thing home and that early bunch of sales in late September taught us various solutions to that problem. Take the example of a muck-spreader bought at Newton-le-Willows over 43 miles from home, a tedious journey to say the least if we had to fetch it ourselves with our own tractor. Fortunately Mary met a dealer there whose big vans went regularly to Ruswarp, our nearest cattle market two miles upstream from Whitby. He agreed to deliver our muck-spreader there for £10. The conversation over that arrangement revealed that he himself had a Vicon Acrobat for sale and he offered to

FARM SALES

leave it at Ruswarp for our inspection. Since we duly bought the Vicon it was a case of killing two birds with but one stone.

Retrieving the large trailer and cultivator was to prove more difficult. We had bought them at Carlton near Helmsley on the south side of the moors. The farmer from whom we bought them had said the trailer was strong enough to carry the cultivator and that help would be available to load it up. So some days later Giles set off with a good supply of ropes to cross and re-cross the moors by tractor. The outward journey was carefree. It is a wonderful way to see the moors, sitting high on a tractor with an unhindered view all round and breathing the wind fresh off the heather. Though he couldn't hear the birds he could see them clearly and was thrilled when he came across a small flock of snow bunting so interested in the road that they let the tractor get within easy viewing distance. The Carlton farmer was as good as his word and with an impressive display of a mixture of brawn and science the heavy cultivator was manoeuvred onto the trailer leaving Giles to lash it down rather in the manner of Gulliver at Lilliput, before setting out for home.

There is no way of crossing the moors without encountering dramatically steep roads. Any attempt to avoid the steepest involves lengthening both the time and the distance of the journey. As there was little prospect of a novice driving fast with a tractor towing a heavily laden trailer it seemed a reasonable gamble to take the shortest route notwithstanding the gradients which after all was the way he had come and so was slightly familiar. How safe or otherwise that return journey was Giles will never know. It was certainly exciting and even frightening when crawling in bottom gear down the hill below Rosedale Chimney famous for its 1 in 3 gradients and its Swiss-style hairpin bends. In such circumstances the driver listens anxiously to the battle between the tractor with its brakes hard on and the free-wheeling trailer trying to take off. Horatius at the Bridge comes to mind – "and those behind cried 'Forward!' And those before cried 'Back!'" Fortunately the tractor won the contest. It was only when Bank House was reached without mishap that the sense of relief proved the degree of strain the event had generated. It was baptism by fire for a farmer of only seven weeks' experience. And yet for any Dalesman born and bred there was nothing remarkable about such a journey: it was all in a day's work for a hill farmer.

Another such "day's work" took us with our tractor and trailer to a sale on the road to Goathland from Egton Bridge, where we bought a hay-turner and a spare, smaller trailer. To come home we had to load the hay-turner onto the stronger trailer and tow the smaller trailer behind that. Our procession of three vehicles had to go down Glaisdale's notorious Limber Hill with its 1 in 3 gradient ending in a sharp hairpin bend to avoid tumbling into the River Esk. It was at that sale that Mary spied an attractive plain oak fire fender she thought would fit our sitting room fire place at Bank House. When giving our address at the pay desk she was delighted to learn that the fender would indeed fit our fireplace since that was where it had come from years previously. It turned out that the vendor's wife was sister to our immediate neighbour.

We soon came to recognise that such coincidences at farm sales were quite common. They illustrated the close-knit relationships binding the wider farming community of the Esk Valley and even of the North York Moors as a whole. For though we were comparative strangers in the district however distant the sale we chose to go to we were almost bound to meet people we knew or recognised as we did at Hutton-le-Hole a few days after our first Christmas. The occasion was not a typical dispersal sale; it dealt almost exclusively with timber. On a farm you can always do with more timber so we had gone just to see if there were any bargains to pick up. As the sale progressed it became obvious that it would be well worth our while buying quite a lot of the wood provided we could get it home without great trouble and expense. We had come by car and most of the timber was in the form of thin planks or laths, much too long for it. While we were hesitating we met two other Glaisdale farmers in the same predicament. Unlike us they had a solution. We agreed that if we all bought quite a lot it would be worth our paying a Glaisdale haulier to bring his empty cattle van. One of them went to find a phone and returned beaming. The cattle van would be with us in under an hour and we were all free to start bidding for whatever we wanted. Before we went to bed that night we had a very useful supply of wood stacked away very neatly under our cartshed roof.

As the years went by we picked up many of the tricks of the trade. Just how much better organised we became can be illustrated by our strategy a few years later when we discovered that there were to be three simultaneous sales far apart, each advertising things we wanted. The day before we toured around inspecting each of them but were left interested in all three, two in particular. The baler we had our eye on was at Gillamoor high on the south edge of the moor towards Helmsley, while a buckrake we wanted was for sale at Staintondale near the coast north of Scarborough nearly forty miles away. Clearly we had to divide our forces but we had only one car. At that time we had three resident helpers of whom one was away so we decided to work in two pairs. Before leaving home we phoned the auctioneer of the

FARM SALES

third sale instructing him to bid for us up to specific amounts, which tactic resulted in our buying a sheep hayrack. Then we all drove to Gillamoor where we dropped off Mary and her assistant before Giles and his helper hurried off towards the coast and reached Staintondale in time to bid for the buckrake, successfully. In the glow of this achievement they rewarded themselves with a diversion to the Saxon crypt of Lastingham church before returning to Gillamoor where they found Mary and her assistant equally chuffed at having bought the baler for a mere £48. It was therefore a very happy crew that drove home over the moors to resume their farming and social activities. Mary disappeared with her fiddle to an orchestra practice in Whitby while the others brought Duchess out of the rain into a calving pen. Let the diary complete the day's story:

> "G returned with M and lantern and Times crossword at 10.45 — kept vigil for an hour – long pause in progress while only two feet visible – finally, taking only three minutes, unaided, Duchess produced a large son backwards! – not breathing – some vigorous work got him going – bed, satisfied, circa 12.40."

## Chapter Eight
## CHRISTMAS, GEESE & BOTTON VILLAGE

Christmas was coming and the geese were getting fat. Or were they? In so far as we had a goose policy it consisted very obviously and simply of preparing for the Christmas market and selling all but the breeding core of Monsieur Mollet and a couple of his wives. We looked forward to having something to sell at last and took it for granted that we should somehow cope with the killing, plucking and dressing when the time came. However, by then our stream of visitors had naturally dried up and our one long-term helper equally naturally was packing to go home for Christmas – "packing" in his case involved finishing making the rabbit hutch he was to take with him. So we were prepared to do everything ourselves. What we hadn't bargained for was a local tradition that expected geese for the table at Christmas to be presented as an exhibition piece of culinary art, highly polished and without the least vestige of feather, all carefully burned off with methylated spirit. The finished article included both the head and the feet though neither would be cooked or eaten. Apparently Glaisdale customers expected their Christmas geese to be set out with severed heads lying forward over their crossed feet with each yellow claw glistening. A bit macabre we thought!

A kind neighbour who came to demonstrate what was wanted gently intimated that our plain, rough and ready standard would hardly do. Fortunately we had killed and plucked three geese at the beginning of the month as a trial run and we had written and dispatched our Christmas Greetings in unusually good time so the two of us were able to give all the hours it took as dogged novices to ensure that our remaining birds might prove up to standard. Some hope! In the event our highly expert Glaisdale butcher protected his reputation by rejecting three ganders while rather charitably buying five geese for a total of £19.58. To achieve this modest return we had reared the birds for nine months and endured two of the hardest days' work so far at Bank House. The diary for December 18th, written as it records at 12.55 a.m., was understandably as short as the day had been long:

*"Plucking all of the time 2.45 – 8 p.m. – except for tea and for M to take Tiberius to vet in Whitby who says he has worms: given injection and worming pills. Later; G a little electrics while M carried on gutting etc."*

That was barely half the story. We resumed next morning, killing, plucking and dressing a further four geese in time to take them to the butcher who was himself still working at 10.20 p.m. We were learning the hard way. The lesson was reinforced two days later when Mary took the ganders to Whitby but could not find a buyer.

Though the two of us had no help with the geese we had not been entirely alone. Upstairs a good friend from London lay in bed recovering from flu. He had come for a short weekend to check our progress in rewiring the whole house, an economy which we would never had dared to undertake without his offer to supervise by means of periodic short visits, supported by guidance delivered over the phone. On his first visit early in November he had set up the project by drawing up a detailed plan of campaign, acquiring all the necessary tools and materials and teaching us how to use them. Alas for him and his wife thick fog persisted throughout their weekend and they had to return south without so much as a glimpse of the dale or its moorland horizon. For all they knew first hand our isolated farmyard might have been in the centre of Middlesbrough. Since then whenever there had been time and energy to spare, Giles had spent many evening hours as a very amateur electrician. Achievement was most noticeable on the rare occasions when bad weather kept us indoors in daylight, but even then progress went at a snail's pace. Fitting a single power socket in the hard sandstone in our bedroom took 1 ¾ of an hour even with the aid of a power drill and cold chisels. In the kitchen we found that the walls were lined with an inch of very hard cement. The December visit was meant to provide a leap forward but that was largely frustrated by the flu. By the time our electrical mentor staggered out of bed we had to drive him to York station.

That trip to York heralded our first Christmas at Bank House. Since there was no need to hurry home immediately we joined in the prevailing mood of Christmas shopping and then, with our baskets overflowing with leeks from the market stalls, attended choral evensong in the Minster. A dignitary escorted us through the rude screen arch to a private box pew among the choir stalls where we could hide our leeks – though not their odour! – and relish the seven part Weelkes anthem unabashed before returning in the dark to our empty farmhouse.

We were on our own at last for the first time since that sunny evening of our arrival back in August. The goose episode was over and the electrics could wait so we had time to take stock; time to reflect with pleasure on our progress so far and to ponder on a wide variety of unanswered questions from the seriously important to the comparatively trivial. Amongst the latter was how to celebrate Christmas – "how should we sing the Lord's song in a strange land?." The most urgent was how to find appropriate resident help for the next stage. Of intermediate urgency were the questions where were we to buy a house cow to save us money and daily trips across the dale and how were we to improve on the recent exhausting and unsatisfactory method of selling our geese? As it turned out a significant part of the solution to all of these lay just over the western horizon at Botton in Danby dale.

Botton village was the brainchild of Karl König, an Austrian doctor who co-founded a school for mentally handicapped children as they were then known. To continue his work he had to flee from Nazism to Britain in 1938. The schools he established in this country were caring for over 600 children by 1955 when he turned his attention to the question of what should become of those handicapped pupils once they left school. The solution took the form of village communities offering mentally handicapped adults lifelong care and security, living in a series of family units which, linked together, formed the greater family of the whole community or "village". The handicapped, known as "villagers", lived and worked alongside roughly equal numbers of carers who were called "co-workers," and everyone shared in common domestic chores, weekday productive work and a rich mixture of cultural, recreational activities. The community was to become largely self-sufficient, but besides providing for many if not most of their own needs, their farms, gardens and workshops produced much to sell to the general public, especially in the locality.

The first of these village communities had opened in 1955 at Botton in the head of Danby dale only a few miles from Glaisdale so that at the time when we first inspected the derelict farm of Bank House in November 1972 Botton village was already a well-established, flourishing and expanding community, fully confident of its unique vocation. Locally, however, it was not much visited nor widely understood, not just because of its geographical seclusion but also because the British public of the day was suspicious of so unconventional a way of life on their doorstep practising organic, biodynamic farming principles and combining the mentally handicapped with carers amongst whom many foreigners were particularly prominent.

Mary had heard about Botton some ten years earlier when she was appointed Personal Assistant to Lady Eve Balfour at the Headquarters of the Soil Association, because her predecessor was leaving to found a new house at Botton. So when we were first wondering if Glaisdale might be at all suitable a place for our venture it naturally occurred to Mary that we should consult her contact at Botton and enquire about the prospects for organic and socially unusual farming in that district. On Giles's very first trip to Bank House we made an appointment to call at Botton and although Mary's original contact had moved on we were met with courtesy and an unexpected degree of interest by a woman who was a founder member and might be called an elder stateswoman of the community. We returned to Abbotsholme distinctly encouraged, knowing that in the unlikely event of our buying Bank House we should have a source of experience and sympathy within easy reach. That encouragement was enhanced soon after when we received a letter from our Botton hostess asking if we had been successful. She concluded:

"It would be very nice if you became our neighbours but I fear not having heard from you that you have not got it. Do come again in any case."

So the first of many services that Botton was to render us was its important contribution to our confidence in deciding to bid for the farm.

There was so much to do on the farm during our early weeks on the farm that neither our energetic holidaying visitors nor our one long-term resident helper, still less we ourselves, ever suggested going out for an evening. Our own little world held sufficient interest and sociability but our spate of visitors dried up as the evenings lengthened and summer gave way to October. We began to notice advertisements, so when we heard that a brass ensemble was to give a public performance at Botton the three of us sallied forth. We had not given much thought to Botton for some time so we were surprised to realise that Botton had not been equally forgetful of us. We were soon recognised and introduced to several people including a woman who seemed to be a sort of personnel manager with a brief for placing both villagers and co-workers. It seemed that we had become the subject of considerable interest. Clearly they had understood the nature of our enterprise, for that evening we were asked if we might be interested in taking in an epileptic who was nevertheless good at hedging and liked to work on his own. A week or so later a farmer from Botton called at Bank House with a couple of friends and an infant, curious to see what we were doing.

Though we said we didn't feel that we were yet ready to take on responsibility for an epileptic, Botton came up with another idea. For about two years they had been looking after a young man who was not technically "mentally handicapped" but whose other handicaps had landed him in the care of the probation officers. We shall call him Dick Bishop. Dick had benefited greatly from his time at Botton but now he needed to move on. Would we consider taking him? We saw his probation officer at length one afternoon in mid November. After a fortnight of deliberation we began to think it was worth trying. The grass was not allowed to grow under our feet. It was now December and Dick's housefather arranged to bring him over to meet us. They toured the farm and stayed to tea. Soon it was agreed that

Dick should come to us for a trial period early in the new year. In the course of negotiations the Botton authorities had learned that we were looking to buy a house cow, whereupon they offered to teach Dick to milk by the time he came to us, something he had not done previously. They had also learned that Mary was particularly interested in getting a dairy shorthorn from a local herd that would be already acclimatised to our conditions. Since Botton had a shorthorn dairy herd they thought they might be able to sell us such a cow. When on the night of our trip to York we went to Botton to see a performance of their "Paradise Play", we were told that they had a cow "Jeannie", that they were expecting to cull but which might suit us instead. We were also invited to return two days later to share in their Christmas Eve ritual. So instead of the two of us spending Christmas Eve on our hillside with our cattle ranging freely under the stars as our principal company, we were plunged into the warm, hospitable atmosphere of Botton's vibrant community life. First we went to Honey Bee Nest Farm to be introduced to Jeannie in her stall among the other cows each of which had a board above her bearing the date of her next expected calving. On Jeannie's board someone had simply chalked the word "slaughter". As it turned out she probably outlived them all and reared more calves than any of the others for although no price had been mentioned as yet, that was the beginning of a fruitful relationship with us that was to last ten years.

From the dim cow byres we were taken to one of the houses to enjoy supper round a large table with the House-parents, their own children and their wider family of villagers. Then we all walked half a mile through the dark to the Community Centre to see a pageant performed which culminated in lighting the candles on a huge Christmas tree. Members of every house had taken jam jars containing candles which they now lit direct from those on the central tree, before carrying them as lanterns back to their own houses, using them in turn to light other candles waiting on their own small Christmas trees. It was a most effective piece of symbolism expressing the relationship of each family with the community as a whole. We got back to Bank House in time to rescue a plum pudding from the Aga, do a little tidying up and set off once more into the night to go to the midnight service in our church in Glaisdale.

In a very short space of time Botton had brought us solutions to three of the questions we had so recently been pondering: our need for long term assistance, our need for a suitable house cow and our appetite for a taste of Christmas ritual. Dick moved in on January 7th followed by Jeannie five days later. And since blessings, like troubles, never come singly, January 9th saw the arrival of a strong and entirely congenial young friend using a chunk of his gap year to see us through to Easter. We shall call him Gerald.

The speed and confidence with which our relationship with Botton developed was not just a matter of chance. It was built on a number of values and convictions held in common, such as the belief in the equal value of every human life and the recognition that everyone however handicapped or apparently lacking in talents, is capable of contributing positively to the common good. Farming at both Botton and Bank House was seen as an expression of man's love of nature and cooperation with it, not as a matter of his mastery over it: it was a means of shouldering man's responsibility for the earth and the plant and animal kingdoms that inhabit it. We all believed also that work should be satisfying in itself and does not necessarily need its reward in payment.

Over the years nearly all our helpers and visitors found Botton inspiring and enjoyed going there. Possibly in some cases a little of their admiration of it rubbed off on us and helped them to appreciate what we were trying to do too. For us at Bank House Botton provided the nearest "health food" supply, our nearest book shop, our nearest printing service and the nearest source of student-aged volunteers with whom our helpers could easily identify, so it was an extraordinary piece of good fortune for us to have Botton village on our doorstep. From the first we regarded it as a community of kindred spirits with whom it was natural and easy to cooperate though it was of course a most unequal cooperation between a village of two to three hundred souls and our tiny household. Nevertheless some of the traffic was two way. From them we gained Dick Bishop: at Botton we met a girl longing for a place on a farm with working horses and she came to Bank House for over a year. Some time later one of our helpers who had got to know Botton during his year with us joined their community afterwards and became a House-father there. On a couple of occasions it suited Botton to have a villager stay with us for a short while: one was a very timid girl who wouldn't go near a cow or answer a telephone. Her most useful achievement was keeping hens out of our strawberry beds, though even that proved too much for her at times.

Botton didn't use artificial fertilisers or buy food from farms relying on chemicals but they were happy to consume produce from us and they became our best poultry customers. A few years after we got Jeanie from them they bought one of our young Devon bull calves to breed from.

Sometimes we resorted to barter. What we eventually "paid" for Jeannie was actually a load of old hay, and it was not unusual to find our diary recording: "Mary dealt with Frazer from Botton who brought frozen food in exchange for chicks", or on another occasion, "D to Botton with eggs to swap and cheese to buy."

For the most part our contacts with Botton were single events arranged spontaneously, but two traditions grew up which proved of such mutual use and enjoyment that they continued over many years. One concerned the production of what became our annual Christmas Greetings newsletter. The first, in 1972, had been an amateur, scruffy effort run off on the duplicator in the Abbotsholme staff room. Its successor was kindly done for us by the vicar of Glaisdale on a similar machine that he used for the parish magazine. Then we discovered the Botton Press and its sophisticated equipment so the 1974 letter was a quality job on quality paper. With a single exception, from that day to this, we have made our pre-Christmas pilgrimage to the Botton Press. It used to involve us in three visits each season. On the first we took the script and discussed any problems; the second was for us to proof-read; the last a day or two later was for us to collect the finished copies. More recently, thanks to remarkable developments in printing technology, the whole process has been telescoped into a single morning provided we arrived with our text safely stored on a floppy disk. It is now a civilised, sociable occasion with time for coffee and an exchange of jokes disguising the serious purpose of our visit until we drive away with our precious booty. Christmas would hardly be Christmas without it.

The other Botton-Bank House tradition brings us back to the beginning of this chapter, to our Christmas geese problem. The experience of our first season was not to be repeated; nor was it! To begin with, in 1974, we bred and fattened fewer geese and we further reduced our work-load by giving up the attempt to meet the local standard of presentation. Instead we looked for less exacting buyers though with less profit. It is hardly surprising that Botton turned out to be the perfect customers for Bank House poultry. In the first place, and most importantly, because they were ready to take their birds unplucked, thus saving us a great deal of tedious work. Our problem was finding time to do the job; theirs was finding suitable occupation for the villagers and apparently plucking fitted the bill perfectly. It didn't take long for a second advantage to become evident, that they didn't need to leave the transaction until just before Christmas. They preferred to buy them early, even in November, and having plucked and dressed them keep them in their big freezers until Christmas. This relieved us of a second problem, that of timing the fattening process so the geese reached their peak of condition in the fourth week of December. Instead our task became flexible. To some extent we could arrange to kill the geese when they were ready and to feed less expensive corn to them once the grass diminished. Nor did we have to kill them all at once. We could keep the smaller birds longer and fatten them more slowly.

For several years Botton bought most of our spare geese as well as a lot of chickens at a time that suited both parties so that we killed them and delivered them still warm for them to pluck. Then it was suggested that the plucking gang of villagers would like to do their work at Bank House. The plucking would be even easier as the birds would be warmer but the main attraction of the change was that it would provide the villagers with an outing, a change of scene and the novelty of seeing another farm, meeting new people and enjoying a different sort of lunch and tea. We first tried it out in 1977 and it proved a great success, so much so that the pattern was repeated yearly until we left the farm some twelve years later. Indeed when our geese failed to hatch out one year and we had none to sell the outing still took place by request because the villagers had come to expect it and looked forward to it. On that occasion instead of plucking the working party was employed fetching in logs from the nutwood, which in truth was better matched to their physical ability and span of concentration.

Usually Mary would spend the whole morning in the kitchen making a great cauldron of soup and baking a lot of fresh wholemeal bread and a big slab of parkin. The rest of us prepared the kitchen to accommodate the invasion for lunch and tea by bringing in a second table and a variety of stools and chairs. The cartshed too had to be got ready for action. Two long planks were put on rows of oil drums as benches and an assortment of big cardboard boxes were placed between them to receive the course feathers and smaller cleaner ones for the precious fine down. We timed the killing of the birds and hanging them from the beams so that they had stopped flapping by the time the Botton dormobile trundled into the yard. One year our friend, the boss of Botton's food store who had arranged the visit, asked us to teach him how to kill geese, so the villagers were taken on a longish walkabout until the lesson was over and the plucking could begin.

It was an extremely sociable experience every year and an unforgettable scene which would have needed a Breughel to do it justice, with anything up to twelve or fourteen people necessarily muffled in layers of overalls, jerseys, scarves and overcoats to keep out the winter cold as we sat through many hours in the doorless cartshed. Sometimes an extra able- bodied co-worker was brought so that with one or two of our resident helpers as well there may have been four or five competent pluckers to accomplish a high proportion of the work. If a villager began to lose heart because of his own slowness he would swap his barely begun bird for one that was well on its way. Feathers floated everywhere yet some fastidious plucker would manage to keep himself largely feather-free at the expense of making barely discernable progress. Others attacked their victims with more vigour than control, tearing the skin in a way that would have made the Glaisdale butcher weep, but after all it was their own Christmas dinner they were mauling. Though a few villagers might remain absolutely silent throughout, there was non-stop banter and gossip from the rest of the Botton party. Naturally most of it was about people and events at Botton so by the end of the day, and even more so after years of repeated visits, we at Bank House began to develop a first hand picture of what it must be like to be handicapped and to be a member of that community.

By lunchtime most of us looked like snowmen. Attempts to shake off the feathers succeeded in warming us up but failed notably to keep feathers out of the house. For the visitors lunch was clearly the highlight of the day. Crammed into the kitchen with a blazing log fire that we soon had to allow to die down, the noise of chatter and laughter coaxed even the previously tongue-tied into mumbling replies to questions designed to draw them out, while the more loquacious could hardly be silenced. Then back to the cartshed which seemed colder than ever, to finish the job. Time went more slowly. The talk subsided. It needed constant close supervision to make sure that the boxes of soft down were kept free of larger feathers with the sharp quills. We must have been pretty successful judging by the comfort and blessed warmth we still enjoy under our duvet some twenty years later. This lasting physical warmth which we owe to those Botton goose plucking parties, now long ago, has always been matched for us by the inner warmth that came from the pervading spirit of loving concern and generous humanity that almost everyone who visits Botton becomes aware of as soon as they set foot in the village. It seems that something of the same spirit was experienced by the handicapped villagers during our plucking parties for whenever we appear at Botton, whether to shop or to attend a concert, we are greeted loudly by name as long lost friends and regaled with details of feathers and parkin or of earning their Christmas dinner, not as if they were remembrance of time long past but as if they had happened only yesterday.

## Chapter Nine
## THE SHEEP

In the Middle Ages when England's prosperity depended on the wool trade and the Lord Chancellor was set symbolically on a great woolsack, most of the population derived their notion of the nature of sheep from some degree of personal experience. This is still the case in the North York Moors for sheep are everywhere. The traditional layout of farms in these dales assumed that every farmer, whether tenant or owner-occupier, would want to keep sheep and need access to the adjacent moorland.

When we came to Glaisdale the majority of neighbouring farmers still maintained a flock but restricted within the confines of their own walls and hedges. More and more were giving up their use of the open moorland though they jealously guarded their right to it. In the case of Bank House however, these rights had been extinguished. It was yet another aspect of our predecessor's damaging neglect of the farm that he failed to register our traditional rights though the powers that be had allowed him seven years in which to do so. As it happened we had no intention of running sheep on the moor for it would have required a measure of expertise and energy beyond our ability to control both sheep and dogs up there. Nevertheless from the very first, keeping sheep on the farm itself was to be an integral part of our plan and our second main source of income.

Sheep have long had a bad press in our urban culture. This has fostered misconceptions as to their character and intelligence whence it is widely believed that they are both stupid and craven. Many instances of their behaviour considered to demonstrate this could equally well be interpreted as illustrating either their independence of mind – they simply are not interested in doing what we humans want them to do – or their faith in the maxim that "discretion is the better part of valour".

The word "silly" was first applied to sheep at a time when the word referred to their defencelessness and vulnerability. It did not imply any lack of intelligence. Anyone who has cared for sheep for any length of time will recognise this basic characteristic that they are prone to an endless variety of ailments and mishaps, so much so that some shepherds believe that sheep are possessed of a positive will to die. At least constant vigilance is essential. We had served an elementary apprenticeship at Abbotsholme and had learned to make it a rule to see all the sheep every day if it were possible. This daily visit was all the more desirable at Bank House because of the state of the farm where they could – and did! – get entangled in barbed wire and brambles, twisting the latter into ropes as they turned and turned, always the same way, in their vain effort to escape. If they avoided this fate they would still fill their fleeces with burdock burrs, get stuck in treacherous bogs, or of course break through a poor bit of hedging. In every case it was important we discovered it sooner rather than later. A gap made by a single adventurer may be blocked in a few minutes with torn off twigs and branches but when it has been enlarged by the rest of the flock that thinks it must be missing something, it becomes a job for posts and rails or netting.

As if these examples are not enough to make a farmer want to see the flock daily there is the curious fact that if a sheep finds itself lying on its back on level ground with its four legs waving in the air it is quite likely to be incapable of turning itself back on its feet again, and if left long enough, it will die. All it needs is a hand to help it roll over.

We had not expected to spend quite as much time as we did our first year or so over such accidents. At least we had reason to hope they would diminish as we rescued the farm from neglect. In contrast was the permanent vulnerability of sheep to foot rot, to combat which we knew in advance would be a very arduous part of our responsibility as shepherds. In terms of evolution sheep and goats were designed for mountain environments not for lowland verdure. In the wild they need rocks to wear down their hooves. By bringing them to our climate and landscape man was taking a gamble and the price we have paid ever since is to have to tend their feet ourselves. This involves periodically paring their hooves which left to themselves would grow longer and longer, curling round underneath their feet, producing discomfort and pain similar to the human suffering from ingrowing toenails.

The cloven hoof, so effective in gaining a purchase on slippery rocks, is not particularly necessary for walking on turf or mud. On the contrary, it becomes a trap for packed, wet earth which is very slow to dry out and harbours a fungus, the sheep equivalent of athlete's foot. This too left untended leads to pain, restricted movement and even in extreme cases to the loss of the foot. Treating the hoof and the fungus is fairly straightforward, requiring a pair of sharp clippers, a can of fungicide and the development of skilful technique. Easier said than done! It doesn't take much imagination to realise why the hard pressed farmer, seeing a few lame sheep or lambs in the next field from the one in which he is working, delays bringing them back from the far end of the farm to the yard of pens where it is easier to catch and control his patients.

## THE SHEEP

One always hoped that the daily inspection would involve nothing more than an uninterrupted stroll, noticing that every animal was grazing or lying down contentedly chewing the cud. But to be sure all was well we had to count otherwise we might miss the casualty hidden in a ditch, or the truant across the beck in a neighbour's field. It is extraordinary how often one has to count, again and again, to arrive at a convincing score. No wonder a sheep-breeding nation uses counting sheep to combat sleeplessness! They may seem still but as for the school photographer, someone is always moving.

Most of the troubles mentioned above occur at any time of year but there are others to look out for at specific seasons. Of these the most unpleasant comes in midsummer with the hot, humid weather, which flies find ideal for laying their eggs in sheep's' fleeces. These eggs hatch into maggots which then begin to eat into the living flesh beneath, a horrible condition known as "fly-strike". On sultry days in particular, before sheep shearing or "clipping" has taken place, if we saw a sheep looking uncomfortable, twitching its tail and repeatedly looking back at its rear end, it was a sure sign of fly-strike and it was urgent to get the victim in to cut away any wet, dung-soiled wool and treat the affected area with strong sheep dip. Spooning off the writhing maggots as they were revealed and chasing them before they escaped into the safety of the nearest bit of dry fleece to begin all over again - ! is something to be taken into consideration by anyone tempted to keep sheep by romantic notions of shepherds and shepherdesses. Fortunately, we were spared this tribulation in 1973 by arriving with our few sheep in August, but we made up for it the following year!

If it is beginning to appear that sheep farming is hardly a doddle, wait for it! We have yet to relate the five major tasks of the shepherding year; lambing, clipping, dipping, the sheep sales and tupping; - major that is measured in terms of importance, prolonged effort and the organisation involved. As the years came around familiarity and experience enabled us to tackle them without losing too much sleep but to begin with we found each of them daunting. We shall deal with each in turn beginning with the sheep sales because they came first even before our opening month at Bank House was quite over.

The sixteen sheep we brought with us from Millholme were clearly no more than a nucleus of the flock needed at Bank House. We knew all along that one of our first important jobs would be to purchase more ewes and to get them in lamb in time for the lambing season of 1974.

Romneys at lambing time

Tupping Time: the daily check

## THE SHEEP

There were many sheep sales in the district through the autumn and fortunately two of these were held in Glaisdale itself, the first early in September and the second a month later. The advantage of patronising one of these was not just the easier and cheaper transport but also more persuasively that they were a family occasion enjoyed by the village as a whole not just by farmers. We should have the chance to meet a great many of them and if we bought sheep from some of them we should be able to keep in touch afterwards and get advice when we needed it.

So after barely a month at Bank House we prepared ourselves for our first Glaisdale sheep fair. We had to decide how many sheep we were to keep altogether, how much we could afford to spend and which breeds to bid for. The first question was simplest. Our Ministry advice suggested Bank House was capable of supporting some sixty ewes in addition to our Devon cattle. The second which was largely a matter of doing our sums in advance, needed quick thinking when our bids were being overtaken once the sheep were in the ring. Much more important and difficult was the question what breed we should acquire. Winsome and her daughters were somewhat mongrel, white-faced and predominantly Romney (Kent) – a fact we had only recently established thanks to the expertise of the Wool Marketing Board graders who could tell the age, sex and breed of any fleece at a glance. As far as we knew there were no other Romney flocks anywhere near Glaisdale. If we wanted to build up a bigger Romney breeding flock we should have to do it gradually from our existing ewes. Ewes we bought in of other local breeds would be for producing lambs for butchers' trade, and since it appeared that local butchers preferred black- or mottle-faced lambs we realised that that was what we should have to produce.

Three days before the first Glaisdale sale Mary went to the regular Malton market to get an idea of current prices and thinking she might buy some store lambs to fatten during the winter for re-sale the following year. However she found store lambs were almost as expensive as ewes so she came back with eight Suffolk-Kerry cross ewes in our trailer. They too could be fattened for resale along with any lambs they might have produced in the meantime. Within two days they were beginning to accept Winsome's leadership.

The sheep fair was always a highlight of the Glaisdale year, socially as well as agriculturally. Several thousand sheep were sorted into several hundred pens set up in a field behind the Angler's Rest which did a roaring trade all day, and whither we repaired for beer and pork pies when were waiting for the sheep we were interested in to take their turn in the ring.

## THE SHEEP

Often they entered with joyous leaps and bounds. They seemed to be saying, "See how fit I am. Buy me!" but in truth it was just to stretch their legs after hours of being packed in pens like sardines. The entry in our diary written that night catches a little of our excitement and exhilaration:

> "Historic day! Fascinating, exhausting occasion. Everything to learn- breeds, names, methods etc  Faces to remember – hurdles and pens-- a technique to copy. How to look at shearlings' teeth."

Amongst the many things we learned about that day was the local custom of "luck money". We were surprised after having bid for some sheep successfully to be approached by the farmer who sold them with some coins in his hand which he offered us in the manner of a tip. Then we noticed similar transactions going on all around us. Though the logic of the practice escaped us we did appreciate the personal contact it entailed with someone who might otherwise remain a total stranger. We made a mental note that when we had stock to sell in future we would make sure we had appropriate money in our pockets.

Something else we were learning was the terminology used in Yorkshire to describe the age, sex and status of sheep which in many cases differed from what we had known in Sussex or Staffordshire. According to the catalogue the thirty-eight animals we bought that day consisted of eighteen "shearling gimmers", ten "gimmer lambs" and ten "ewes". This meant that we had eighteen ladies in the prime of life, ten young ones born that year and ten dames that we had judged to have enough teeth to see them through one more lambing season. Of these last no promises were made. In the delightful phraseology of auctioneers "you take them as you find them".

A carter from the village delivered our purchases to our farmyard and departed, whereupon the nervous newcomers demonstrated their mettle by immediately clearing the cattle grid. If we hadn't known it before we knew it now: Mashams are wild rovers compared to law-abiding Romneys. So our first Glaisdale sheep sale day ended with us retrieving our Mashams by running around our neighbour's field as human sheep dogs.

What a day! There couldn't be another like it. Nor was there, for the following September we ourselves had lambs to sell. We avoided the expense of a carter by walking them to the sale down the two miles of road, for by then we had our own dog and what we lacked in skilful control we made up for with the enthusiasm of young helpers and by linking up part way with a neighbour and his posse of sale sheep. But however much experience you acquire you cannot escape the fluctuations of the market. In

# THE SHEEP

1973 when we had sheep to buy but none to sell, prices were unusually high. In 1974 when we went as sellers but not buyers, prices had fallen disastrously and our whole year's crop of lambs sold for under £300. We should have been more cheerful if we could have foreseen that in 1977 when the market had bounced back, our sale of lambs would bring us in more than four times that amount. But that is farming!

The shepherd's year is shaped by the gestation period of his charges which is close to twenty one weeks, so if a ewe is served on Guy Fawkes' night she is likely to lamb about the following April Fool's Day. Most sheep farmers tend to stick fairly near to the same day each year so the rhythm of their ewes' lives allows them time to recover from one year's lambing and build up strength for the next. The gentlemen of course have other ideas and have to be kept securely away until the appointed time.

The terms "ram" and "tup" are generally interchangeable but as an alternative to "tupping" the word "ramming" simply will not do to describe the period when lambs are begotten. Southern visitors to Bank House talked about "rams" in the "autumn" but come the "back end" it had to be "tupping time". There is some latitude to farmers in choosing when they want their ewes to lamb and having made that choice a simple calculation gives them the date for tupping to begin, the day when the previously separated tups are put with the ewes. What influences this decision most is their expectation of the weather when the lambs are to be born coupled with the likely amount of grass available by then. For our first lambing season at Glaisdale we were planning to begin early in March, but first we had to buy a tup. What breed should we look for?

All the sheep we brought with us from Millholme were Winsome's offspring fathered by either the Suffolk tup which happened to be in the next field – with which Giles indulged in shoving contests, head to head! – or Tristram, a lugubrious Dorset Horn tup we bought under the misapprehension that the genes accounting for the horns on Winsome's twin brother and on a few of her progeny might have come from Dorset. Since then we had discovered Winsome's predominantly Romney blood and as we were evidently happy with her growth and temperament we were inclined to get a Romney tup. Further factors favouring that choice were the breed's reputation for placidity and sound feet, distinct assets for Bank House with its poorly drained land and untrustworthy defences. The tup we were about to buy would of course serve all our recently acquired Masham and Suffolk ewes as well as Winsome's brigade but the overriding consideration was to build up our Romney breeding flock so Mary began

## THE SHEEP

negotiating for a tup from distant Kent. Fortunately we discovered that a well-known Lincolnshire farmer was having some Romneys delivered at the right time for a tup for us to be put in with them, so instead of having to drive all the way to Kent Mary set off for Lincolnshire with our little sheep trailer getting there and back in a day. She returned in the dark with the tup which we christened Ethelbert after the ninth century king of Kent. His initial behaviour was hardly regal. He took twenty-five minutes to pluck up courage to come out of the trailer and was happy to be shut up on his own for the night in a stable. The next morning he submitted tamely to being daubed with raddle before being put with a party of selected ewes, of whom he had served only five after four days. We took it to be an indication of what animals suffer when transported great distances. Thereafter he took charge of his kingdom and proved himself able to defend it, as we shall see shortly.

Tupping posed us a challenge of an unusual sort. In order for us to know when to expect each ewe to lamb in the spring we had to know when she had been served in the autumn. We adopted the traditional practice of putting raddle on Ethelbert so each ewe he mounted was marked with a red posterior. For that and the following year or two we used the cheap, laborious method of mixing red ochre with oil and smearing it with a stick on his chest, a job that needed repeating every few days especially in wet weather. We had to change the colour used every three weeks otherwise we shouldn't have noticed if a ewe was served a second time. This happened more in later years with those served on the first day when in frenzied excitement the tups would dash about mounting ewes that were not in season. It became easier a few years later when we caught up with the times and bought long-lasting crayons held in position with a leather harness. Then we had to inspect the flock carefully every day and record which ewes had been served since the day before. To be able to do this with any certainty we had to be able to identify each individual ewe, no easy task since we had recently acquired 38 anonymous immigrants which at first, like legendary Chinamen, looked indistinguishable. Mary spent a great many hours on a crash course of identification, an essential, helpful part of which involved giving each ewe a name. Amazingly Mary soon knew every one of them by sight and could name it.

It was not the increasing number of our ewes that necessitated our having a second tup the following year so much as the fact that in this district white-faced lambs consistently fetched a pound or so less in the ring than lambs of similar quality with black or mottled faces. So in 1974 the remedy took the form of Wykeham, our first Suffolk tup, so called after we scoured an atlas for names from the right county. Alas he was not fated to prove the wisdom of our new policy as the diary for October 24th tells us:

> *Shades of last March – another tragedy of Mother Nature. Wykeham killed by Ethelbert for penetrating his territory. Somehow the Horse Pasture / Square Field gate was down and Wykeham and his flock invaded Square Field. A primeval battle unseen by human eye, necessitated by ram law, resulted in Romney destroying Suffolk. Mary rang Egton slaughterhouse – nothing doing – so she spent rest of the day skinning, gutting and disposing –on the open trailer in the yard.*

The unfamiliar strong taste of mutton takes a lot of acquiring and that winter, our second at Bank House, there was an awful lot of mutton for us to eat our way through!

The basic pattern of our lambing procedure did not change radically over our eighteen years at Bank House. Every season had its own flavour largely determined by different weather, different helpers, the improvement of equipment and accommodation, and of course our growing experience. From the start we recognised the great importance of saving as much grass as possible at the near end of the farm as vital food for the ewes just before and after lambing and of course for the lambs too once they began to eat grass themselves. To achieve this we had to keep the flock from early December until the beginning of March at the far end of the farm even though it added greatly to the task of looking after them, especially after Christmas when it came to taking the daily rations of hay to which concentrates were added as lambing time approached. More grass was conserved by delaying bringing up each lot of expectant ewes to the farmyard until shortly before they were due to lamb. Each batch progressed through the same series of stages. First they were brought to the bank above the yard where we could watch them closely, even from our bedroom window. Every evening we brought them into the yard to be fed. Any ewe we judged to be imminent was kept in the yard and visited during the night: the rest returned to the bank where if we had made a mistake they could, and occasionally did, wander away to lamb in a secret corner some hundred feet above us. As each ewe lambed she was put in a pen in the sheep barn or the cart shed where she stayed for a few days. Besides its door to the yard the sheep barn opened into the Garth, a small field of about three quarters of an acre which acted as a nursery for generations of lambs to experiment with their legs and teeth. When the next ewes to lamb needed the barn and the Garth the first batch were taken to the nearest fields below the yard for the day and after that, weather permitting, they were launched into the

# THE SHEEP

wide world, usually Holey Field with hedges providing shelter from winds from every direction, but not before we had put small rubber rings on every lamb's tail and on the wether lambs' testicles.

In all this there was a great deal of valuable work any novice could do; helping drive the sheep at every move, tying hurdles to make pens, filling water buckets in each one and seeing every ewe had some hay and holding lambs while rubber rings were being put on. Even the curiosity of very young visitors in wanting to see a lamb born was useful and saved us watching ourselves. We also owed hours of unbroken sleep to our long-term senior helpers who took turns in carrying out night inspections: these we developed into a system of relay reports so that unless they found something serious enough to warrant waking one of us they would leave written word for the next inspection. Some scraps of paper survive with their barely legible messages for the next bleary-eyed watch. One helper turning out at 3.00. a.m. found the following account:

> MIDNIGHT. *Nothing new. Foison fine. Felicity's foster sucked for 20 mins. She (F) was very flustered at times but quietened down when I sang a bit. Every time (i.e. not just coincidence). Eventually all was peace but I've left her tied up. Hermia's two had three shortish sucks but were very tired and kept dropping off.*
>
> *Other lambs in barn were having riotous games of bucking bronco and leapfrog over the mothers so I told them all to stand in the corner till they learned respect...nothing in the Garth of interest.*

A normal birth is one of the wonders of nature at which you never cease to marvel, especially if it takes place under the stars. Charmian was lying easily chewing the cud when you saw her, high up the Garth at three in the morning. It is a fine night. She looks at you without making any comment. You decide to leave her and return to bed. When you see her next six hours later in broad daylight she is standing so that two lambs with ecstatic tails can fill their little stomachs with warm milk containing vital colostrum. They have found it themselves. Perhaps you needn't have hauled yourself out of bed. After all, there had been nothing to do but watch and give a word of encouragement. If you had been there at five you might have seen her getting up and down again and again unable to find a comfortable position, or a little later putting her head back and straining. Then if you had looked closely you might have seen an opaque bubble appear, to be burst from within by a pair of whitish hooves. If so you would not have worried or interfered. In a few moments a little sack of life would have slithered out with its entire head and body in an envelope of mucus.

# THE SHEEP

Charmian would have turned her head to see what the commotion was and then coming to her senses would have scrambled to her feet and begun licking furiously with flickering tongue, freeing the nose and the mouth so the lamb could breathe. Not long afterwards she would have gone through it all again with the second lamb.

In Charmian's case you needn't have got up, but not all births are uncomplicated and sheep's vulnerability is never illustrated more vividly than when lambing. There was the case of Daphne and Deirdre who lambed near each other on a little hillock. By the time we arrived there were four lambs one of which had rolled away and was yelling. Both ewes were upset and seemed to be claiming it. We put them all in pens in the cartshed, each ewe with two lambs. Returning later we found Daphne was still in distress, calling to Deirdre's pen, but we were taken aback to see Deirdre licking another newly born lamb. One of them had triplets and it was obvious which. The puzzle only existed because we had not been there earlier.

Of course there were tragedies as well as happy endings. Our lambing baptism was hardly propitious. A week before any lambs were due we found Charity down by the beck with two live lambs and one dead. What made the losses particularly distressing was that so often we had not been there but stumbled on corpses unexpectedly. We had to learn to stop our minds running on the lines of "if only!".

Every birth is a drama but we were so struck with the case of Despina that we described it in full in our following Christmas letter. We quote it again though it happened years after our first season because it illustrates so many aspects of our life as apprentice shepherds.

*"Despina had not lambed before and demonstrated her inexperience and her wayward nature by selecting a spot one foot from the edge of a precipice, half an hour before dusk, with rain threatening. From down here in the yard her form, silhouetted on the skyline, was seen to ignore our distant summons. The reason guessed, the rescue undertaken. Fortunately the lamb protested at being carried jerkily down the face of the bank and Despina followed in spasmodic fashion until she baulked on the steepest slope and tried in vain to deliver herself of a second lamb, all head and no legs. To force a head back in and fish for front legs in order to help out a straightened lamb while holding the mother still, is awkward at the best of times and places without having to prevent another wriggling lamb rolling to destruction. All one could do was shout for help, hold onto Mum and warm the first-born in one's coat until assistance came with many hands and a basket of Aga-warmed jerseys. What was finally extracted resembled*

# THE SHEEP

*nothing so much as a glove puppet: large, lolling skull and empty lambskin rag which inflated with mouth to mouth blowing. One flicker of life was followed by limp stillness, then another flicker, massage and more mouth to mouth, warm dry against chill wet, and then a splutter and a breath. Never had we spent so long witnessing the knife edge balance between life and death, or rather between birth and non-birth.*

That was just the beginning. The diary for March 18th carries on the story succinctly:

*M & G spent evening trying to save Despina's number 2 – gave it tiny bit of Mum's colostrum and some of Fiesole's milk via tube and funnel – too weak to suck – lamb warmed in Aga, then in front of fire, then in basket in D's pen till next feed about 10.30pm and 2.30am and 8.30am.*

By way of an epilogue the diary for March 20th reported:

*M had further success with Despina's Lazarina which was seen to suck on its own.*

During lambing time we had lived at close quarters to our sheep night and day. Then suddenly they were gone, banished as it were for the whole summer and largely out of sight and earshot from the farmhouse except for the last two sheep events of our first year, clipping and dipping.

Though we should have liked to be able to shear our own flock it was never seriously on. We had picked up some pairs of traditional hand shears at farm sales and very occasionally used them to remove a whole fleece if, for example, a ewe had died. Their regular purpose was for "dagging" – the expressive local term for removing the mucky wool round the sheep's hind quarters. Electrically-powered shearing machines were expensive and daunting too for beginners. We should never have got beyond the novice stage with such a small flock, getting practice only once a year followed by a long gap in which to lose whatever knack we had acquired. In any case even with paid expert shearers from outside there was still enough for us and our helpers to do.

The scene was set in the sheep barn, freshly littered with a thick carpet of wilted nettles and thistles to prevent the straw in the recently vacated and dismantled lambing pens from spoiling the fleeces. It's the devil of a job extracting fragments of straw from wool. Besides we wanted to remove all the nettles from round the yard and buildings at least once a year.

Many shearers preferred to work in pairs and often brought their own assistant shearer with them. Once a young assistant let a ewe escape before he had finished with her. She had obviously had her eye on the open door for she shot through it and away up the yard, half naked, trailing much of her fleece and chased by a barking dog. She was recaptured and returned to the bashful shearer. "I trusted her too much", he explained !

The shearers worked away without a rest apart from momentary breaks to straighten up between sheep. We had two main jobs to do if the whole operation was to go smoothly and not waste a shearer's time. Some of us were responsible for producing a constant supply of ewes as soon as they were wanted. When one fleece fell to the ground and its erstwhile inmate had scarpered off, another had to be ready. Anyone doing this had to be strong and determined enough to grab the next victim, lug it across to the shearer and suddenly (preferably) upend it to a sitting position which in most cases ended the struggle. This was exhausting work on a hot day, and it usually was a hot day because fleeces had to be dry. A heavy shower could postpone the whole operation, possibly for days.

Our other job was to deal with the fleeces. Since fleeces grow from the inside, the outsides looked tired if not actually dirty after a year's exposure. It is a delightful surprise the first time you see the almost dazzling cleanness of the newest wool that had been hidden close to the body now revealed by the shearer. We had to gather up the fleeces out of the shearer's way and spread them out flat, outside uppermost. With practice this could be done with a single skilful flick; it could also take a long time if bits of it got wrong side up. Before it could be rolled up all the bits of dung sheep cannot avoid dropping on themselves have to be cut off. The final bundle was held together with a rope made by twisting the wool at the neck end, then it could be packed in one of the huge hessian sacks, hung from the beams, provided by the Wool Marketing Board. By the end of a long afternoon's rolling up fleeces one's hands were marvellously soft from the lanoline in the wool. Then we sat down to recover, but the shearers usually went straight on to the next farm where another flock was waiting.

Sleep was not easy on the nights after sheep shearing. Lambs know their mothers by smell much more than by voice or appearance but most of the comforting familiar smell was removed with the fleece so when the shorn ewes were released and went calling for their offspring they weren't recognised. It was heart rending to watch desperate, yelling lambs walk straight past their mums and refuse to listen to what they were saying. For quite a while the cacophony produced by ewes and lambs together was dreadful. It died down gradually but it took all the rest of the day and well on into the night for all the families to be reunited. Nevertheless sheep

## THE SHEEP

shearing brought a great sense of release from the knowledge that our ewes were no longer sweltering inside their inescapable blankets and free from immediate danger of fly-strike.

There must always be a pause between clipping and dipping while the sheep grow a new coat, otherwise there wouldn't be enough wool to absorb the strong-smelling fly-deterrent dip. That year we needed that pause to be considerable if we were to find time during the haymaking season to prepare for dipping. Bank House had not always been poorly farmed, indeed under its last tenant farmer up to 1952 it had been maintained exceptionally well. At some time in the past dipping accommodation had been built that was the envy of the dale. Besides the essential deep bath with a well next to it for someone to stand in close to the sheep in the bath he was dipping, there was a fair-sized collecting pen with a little gate through which each animal in turn was ejected straight into the chemical brew. There were steps up out of it to a small dripping pen where the dipped sheep stood streaming on their escape from the bath. It sloped so that most of the surplus dip the sheep took with them flowed directly back into the bath, an essential economy as well as a damage limitation for the land.

Alas, empires are not alone in being subject to decline and fall, as the splendid Bank House dipping facilities witnessed. Having no sheep our predecessor never used the sheep dip for its proper purpose and preferring to avoid any unnecessary effort he used the whole dipping area for a convenient receptacle for rubbish of all sorts. When we bought the farm the bath was full to the brim and overflowing with scrap iron, tins and broken glass, tangled with odd hedge cuttings all of which were submerged in rainwater because the only drain was blocked. To empty rainwater or the poisonous dipping mixture itself we had to use a length of hosepipe and set a siphon going by sucking it hard and stopping just in time! For the rest natural decay had done its work unchecked for at least nine years. The wooden partitions needed a great deal of rebuilding, the gates needed new hinges and latches, the concrete lining the bath and the floor of the pens had deep cracks which had to be weeded and then patched with new cement. Fortunately the dip was conveniently close to the farmhouse so much of the work or clearing up and restoration was carried out in those spare half hours when a promised delivery of straw or tiles hasn't yet arrived or when lunch isn't ready. Though we didn't finally dip our sheep until August 9th our diary records as early as June 25th that: *"we started long job of getting sheep dip ready: weeding, burning rubbish, sickling and baling out sludge"*. We had been busy on and off ever since.

# THE SHEEP

As a useful last preparation, on the morning of the day we were dipping Giles went across the dale to help a neighbour dip his sheep and to observe some techniques and pick up some tips. In the afternoon after all those weeks of making ready we dipped all our sheep, over a hundred of them, in only two and a half hours. Giles's parents were agog to know how it had all gone. Mary wrote to them:

> "What is usually a rough and unpleasant business was done beautifully, calmly and peacefully". Her biased judgement was corroborated by a visitor after watching with close interest. "Throughout I could feel that sense of trust and patient sufferance the creatures adopted".

For the best evidence of how sheep on any farm are treated go and watch them being dipped.

We had got over another hurdle of the farm year. In future it would never be so forbidding a prospect though we were to dip much greater numbers. This increase was because we invited our neighbours to join forces and bring their flock – much bigger than ours – to be dipped at Bank House since they had no equivalent facilities. Sometimes we couldn't deal with so many sheep in one day but the extra hours it took were sociable and highly educative. We learned a vast amount of local history, farming techniques and traditions from the non-stop gossip and banter that distracted us from our aching backs. One thing Mary taught them in return, that dipping is much more pleasant when you are working in a bower of flowering honeysuckle and roses. Sheep crowded for long in pens leave a great deal of muck behind them: more so when apprehensive. No other shrubs on the farm flourished so spectacularly as those around the sheep dip.

The most unforgettable moment of our years of sheep dipping had little to do with sheep or dipping apart from setting the scene. It was one of the hottest days of summer. Giles stood ankle deep in foul water in the well having just lost the struggle to dip sheep but not himself, when a voice from another world inquired after a Mr Heron. The speaker looked and sounded like someone auditioning for a part in "Yes, Minister". He was indeed in pinstripes and carried a briefcase. He disguised his surprise on discovering that the figure in thin shorts and ragged shirt with his toes protruding from old school plimsolls could also be described as one of the proprietors. Could he have a word, please? While the others welcomed a respite Giles and the stranger paced the yard together, polished shoes and squelching plimsolls step for step. A certain young man had applied for a job in a Ministry – which could well have been MI5. Had he been living at Bank House between the dates he had given? Had all his time been occupied on the

# THE SHEEP

farm? Had he really worked without pay? How was it that a university graduate in Russian came to be doing menial farm work in such a remote spot? Had he applied for a job? Why had he left? Did we know anything about his, er, background and, er, character which might make him vulnerable to anyone who wished to be, er, awkward? Giles began to wonder if he would ask if any particular interest had been shown in the early warning system at Fylingdales visible from the top of the moor behind us. Instead the visitor uttered polite thanks and drove off. Giles splashed back to continue dipping. Tony got his job.

## Chapter Ten
## AT FULL STRETCH:   OUR FIRST CALVING SEASON

In many respects the calving season mirrors that of lambing and in our first year they coincided as well though we had not planned it that way. While we timed our tupping so that the ewes would lamb in March we had arranged that our original eighteen Devon heifers should run with bulls on each of their farms before they came to us so as to calve in February 1974. This plan was to prove no exception to Burns' observation that "the best laid schemes o' mice an men gang aft agley". As it turned out not one of our Devon heifers was to calve in the month designated.

In the first place we were completely outmanoeuvred by Duchess. At 7.20 a.m. on September 22$^{nd}$ we were puzzled by a distant brown heap that had appeared in Wood End. Binoculars revealed it to be a wet calf and by the time we got there it was already standing and sucking. As she was our first lady we called her Eve. In human terms Duchess was a teenage mum. How she got herself in calf months before we had even arranged to buy her we never found out. It remained her secret. During the weeks following Eve's arrival we scrutinised the other heifers daily with greater care than ever looking for any signs of swelling udders, those tell-tale indications of approaching birth, but we looked in vain. Gradually, with diminishing expectation, we settled down to "wintering in the open", waiting for February.

We knew from the start that in our first season most calves would be born out of doors because the few existing buildings would be bespoke for a sequence of expected newcomers, Jeannie, two sows and a bull, always reserving shelter for casualties. Besides, even our most optimistic calculations didn't suggest our new buildings would be ready for occupation until our second winter. Meanwhile before the coming calving season we were engaged in a race against time as far as fencing was concerned.

After the exodus of some of our heifers into the neighbouring woods at the beginning of December our fencing operations naturally concentrated on the farm boundaries. It was only in mid-January that we felt able to re-direct our attack by erecting a new fence all along the nearside of the nutwood. Until its completion on February 22$^{nd}$, the herd we wanted near enough for us to hay twice daily from the Dutch barn in the garth could range freely over a third of the farm, an area including the three top fields, the face of the main bank, twelve acres of the nutwood and another five fields beyond it. This made supervision of any heifers we thought

## AT FULL STRETCH: OUR FIRST CALVING SEASON

particularly close to calving a chancy business as well as a tiring one, especially in the dark by the light of torches and a lantern. So it came about that on January 19th we failed to find Edna in time to help her. She had chosen a perfect little shelter in the bracken but for some reason the calf was dead when we found it. There was no time for grieving. Speed was essential because the cost of a foster calf would be a great deal less than that of the yearling it would grow into provided we could get Edna to suckle it before her milk began to diminish. After much phoning around we discovered there was a four week old bull calf available for fostering on the south side of the moors. By evening we had bought it and fetched it to Bank House. Next morning Mary succeeded in getting it to suck from Edna while she was tied up. Ten days and a lot of work later Edna had accepted it as her own though we called it "Perkin" after Perkin Warbeck the 15th Century Pretender. We had succeeded with a vital procedure we had only followed previously with sheep. We were learning.

Then it was Poppy's turn. On the morning of the last day of January she failed to turn up with all the other heifers for the hay we put over the garth wall twice daily. We found her eventually, a third of a mile away just beyond the nutwood, and a sturdy bull calf was lying snug alongside her. She had chosen a sheltered hollow in the bracken high up in Second Cow Banks, conveniently near to a tumbling stream. There was little prospect of our walking them back to the farmyard, still less of our carrying such a heavy, new-born calf so far over such terrain, and as they were out of reach of our tractor we decided to leave them and hay Poppy where she was. When we brought more hay in the evening the calf was sucking competently. There was no change on Friday. Saturday's diary recorded: "*Poppy and calf hardly moved from the cradle in the bracken all day,*" though it also mentioned "*foul weather!*". On the Sunday we found the "cradle" deserted. Poppy had rejoined the herd for haytime but was happy to follow us back to the nutwood in the wake of a private haybag. She led us straight to the calf lying obediently where she had told it to stay in the better shelter of the wood itself. On the Monday the calf, which we now called De Quincey, got himself stuck in one of our predecessor's pheasant traps in the middle of the wood. "*M released it at Poppy's request,*" says the diary. By Tuesday Poppy was eating regularly with the others but late in the evening rain changed to snow so we were a little anxious to see how De Quincey was faring on Wednesday morning. We needn't have worried. At length we found him where the thick trunk of a fallen tree arched over a hollow at the near edge of the wood, almost within sight of the farmyard. Who could tell what intelligence led a six day old calf in the snow to the perfect natural shelter?

## AT FULL STRETCH: OUR FIRST CALVING SEASON

For years until the tree trunk rotted away, "De Quincey's Bower" was one of the most popular attractions for visitors on a tour of the farm.

Just before Poppy produced De Quincey we added six more in-calf heifers to our herd, crosses half Devon and part Hereford and Friesian. At less than half the price of our pedigree Devons we bought them as a cheap, temporary boost to our income as we intended to sell all their offspring and not keep their daughters as part of our permanent breeding stock. Their mid-winter arrival was memorable. On a late January night of torrential rain our household of four had gone to a concert in Guisborough. We were only two miles from home on our return journey when we found the road blocked by a big cattle van that had broken down, halfway up a steep hill. We stopped to see if we could help, only to be asked by the driver if we could tell him the way to a Bank House Farm, Glaisdale! "Follow us!" we said. He had had a terrible journey with his van breaking down repeatedly. That it couldn't cope with our hills we saw for ourselves when his engine failed four more times in the remaining two miles. As the diary reminds us:

> *"eventually got back about midnight – six heifers let out in yard. Two pigs and storage barrels left for night in van. Driver had bacon and eggs. All had soup. Bed 2 a.m. after many yarns."*

So we entered February our expected month of calving. Much of our time and energy went on feeding and watching sheep and cattle wherever they were. Part of our difficulty was that we didn't know our heifers' individualities: how their bodies and temperaments would react to calving for the first time. Plum in particular kept us guessing. She seemed to be beginning to "bag up" week after week and we kept on visiting her – to no avail. In later years we knew only too well how huge her udder became before she calved and how her teats would swell until her new born calf couldn't take hold of them to suck out the milk. The only increase in our herd in the whole month of February came on its very last day with the arrival of Barton Bonus Esquire, our first bull – of which more anon. He merely complicated our calving by giving us more to do and occupying one of the few loose boxes. What affected our calving arrangements more positively was that on February 22nd we at last finished fencing off the nutwood from the near end of the farm so halving the area we had to patrol in search of missing ewes and heifers. The Poppy story would not be repeated.

March entered with six inches of snow, very much the lion rather than the lamb, the month of Mars the God of war. In retrospect this came to seem entirely appropriate because we were about to enter on the toughest

## AT FULL STRETCH: OUR FIRST CALVING SEASON

period of our Bank House venture. So many things went against us when there had been good reason to hope that they would go our way that it became difficult not to entertain superstitious notions that malign stars or those capricious Greek Gods were to blame. At our lowest time three heifers died in the course of calving within a period of only four days, and another died a short while after. Remarkably two of the four calves involved survived but three others calves died after having seemed well established, one of them from a poisoned tongue. The bald statistics can give little idea what those days were like. To mention each tragedy, let alone describe it, would be unprofitably harrowing and give them an exaggerated importance compared with so much good fortune we enjoyed but perhaps one example may be permitted on the principle of "warts and all".

Valetta was one of the cross-bred heifers from Sussex. We found her collapsed on a slope above the lower nutwood having started to calve but temporarily given up. A pair of hooves were visible: she must have been at it for some time. We couldn't shift it so Mary phoned our neighbours and then the vet. Jem arrived with cords and put a slip knot around each ankle so he and Giles had something firm to grip while pulling together with all their weight. There was a dangerous moment when the calf stuck, half way out, making breathing impossible, then out it slithered. "Blow, Mr Heron! Blow down its throat!" Jem shouted, and for the first time in his life Giles found himself trying to give mouth to mouth resuscitation, to an inert calf. Miraculously it worked, but Valetta didn't stir. The vet arrived, gave Valetta several injections and left us with instructions and further drugs to administer.

We called the calf Malteser because of the large, round chocolate drop patch in the middle of his forehead, and we took him into the kitchen to warm by the Aga on straw. It took a long time teaching him to suck from a bottle while he was played with by Tiberius, our adolescent kitten.

A long, wet week of nursing ensued but Valetta never stood up again. We took up bales of straw to give her something to lean her head on. Others were placed below to stop her slipping downhill, but her sister heifers found them and started to eat the straw, so we made a stockade with fencing posts and laths to keep them out. We took it in turns to carry up titbits of food and buckets of water to tempt her to eat and drink. We built a canopy with spars and tarpaulins to keep off the worst of the rain. Sometimes there was a glimmer of hope as she nibbled something or moved a little but it was ever more obviously a losing battle, especially when she groaned. Her death at the end of a dismal week was made even more

## AT FULL STRETCH: OUR FIRST CALVING SEASON

depressing when we had to bury her on the rocky hillside where she lay and where with our help she had given birth to a strapping son though she didn't know it.

There was so much else for us to do during those weeks, quite apart from our calving responsibilities, that we should have been more than fully occupied even if we had had no cattle at all. In the first place of course there were all the kindred calls of lambing and the two often competed. We lost more than one lamb when ignoring a ewe we knew was lambing because we were preoccupied with a protracted calving. But there was much more besides. Very little of our normal routine work on the farm stopped for the duration of lambing and calving. As long as sheep kept on being caught in brambles there was urgency behind our clearance programme. As soon as the earliest lambs and their mothers were put out into the lower fields, each day revealed gaps in hedges and fences that couldn't wait to be mended. Then there were the two fields waiting to be ploughed, cultivated and sown in time to catch the growing season. Ever since frost had made it possible to begin ploughing in the first days of January, it had been too wet to continue so the weather that improved over the last ten days of calving also imposed hour upon hour of tractor work in the field.

Another two claims on our time and energy in those weeks concerned accommodation for us humans and for our pigs. The former did not directly provide work for us to do but for the local builders we had commissioned a year previously to re-roof the farmhouse and carry out a lot of badly needed repairs and alterations internally. They erected the necessary scaffolding at the end of November but didn't turn up to start work until the first full week of lambing and calving. We couldn't put them off because we needed the grant re-roofing would bring. Dust, noise, extra pots of tea and frequent consultations to make decisions, were the order of each day indoors while outside the yard became an uncontrolled car park that made animal movements hazardous. In the midst of all the chaos, with tiles being slithered off the roof down a chute, sat Mary's 80-year-old artist aunt, painting at her easel, happily oblivious of everyone else. A lady of regular habits, she suffered our erratic meal times with great patience and would lay the table in hope at the appropriate time.

And the pigs? Since their arrival on that memorably wet January night our two black sows had been occupying a stall in a stable we would have liked to use for lambs and calves. We started to build them a permanent "house" in the nutwood as early as March 16th, but progress was slow and occasional; slow because rocks made it extremely difficult to set in corner

## AT FULL STRETCH: OUR FIRST CALVING SEASON

posts at all square; occasional because we could not spare time to work on it every day and even then we were frequently called away on nursing or midwife duty. Valetta lay in her stockade only 40 yards or so above the site. But we persevered. On April 1st the roof went on and the next day the sows took possession most appreciatively.

With the new month Mars seemed to be losing his grip. The weather improved and our calving fortune altered dramatically for the better. The last six heifers to calve all managed without harm to themselves, five of them entirely unaided. Their six healthy calves did much to restore our flagging if not shattered confidence and enabled us to reflect on the whole episode with something approaching an open mind. We had learned a huge amount and there were many more lessons for us to learn but we didn't think they proved we had to make basic alterations in our overall strategy. We had suffered grievously from our assumption that the barnful of hay we had inherited would provide suitable nourishment for in-calf heifers. That couldn't be repeated. Nor would the conscious risk we had taken by stocking the farm with cattle before we had appropriate buildings for them. We had been pushing on with preparations for our stockyards ever since we had bought the farm. Surely something substantial would be up for next year's calving? By that time, too, instead of first-time heifers it would be experienced cows calving.

There was nothing all that unusual about most of the items that crowded our timetable. Most of them constitute the ordinary life of farmers in these parts, especially in the spring. What made our lives so intense and exhausting besides our inexperience was the way everything coincided and what seemed a persistent streak of bad luck as with the calf that was helped out by the vet with its heart still beating but it couldn't be got to breathe. A disappointment that was undeniably due to chance set back our planned building up of the number of our pedigree Devon breeding stock more or less by a year. At the end of our first season of calving Duchess's Eve was the only additional pure Devon heifer. All the rest were bull calves.

The sequence of events during that calamitous first Spring season was recorded in remarkable detail in our daily farm diary which stuck for the most part to bald facts. For evidence of our personal reactions to those events we have to turn to some letters which we wrote at Easter once it was all over. There was no denying the strain we had been under, as Mary frankly admitted:

*"I am not sure that I want to live through another such nervously exhausting six weeks – our Lent – with fresh calamities more than once a*

day, but this sunny calm Easter Day we are enjoying a corresponding peace and already the harrowing episodes are slipping into the past. All the orphans are happy with their foster parents."

She went on to count our blessings:

"the newly calved heifers are delightfully friendly – several let me milk them now and give enough for their calves and some for our visitors. We have sixty six enchanting lambs and most of the wild sheep we bought in September now eat out of my hand which will be an enormous boon next year."

Giles related our recent farming experience to Easter. Referring to the loss of so many animals and especially to the heifers, he wrote:

"Now it seems more in the nature of a sacrifice. Somehow the Easter story has got translated into our farming legend and I find myself regarding the four dead heifers as if they have bought our survival at their own expense."

## Chapter Eleven
## BLACK CARR BUILDINGS

There is a field at Bank House long known as "Black Carr". According to the dictionary a carr is "wet, boggy ground; a meadow recovered from the bog". In this case the term is doubly apt as there is a great deal of black soil and wherever you dig in that saturated ground you uncover evidence of successive generations of drainage; evidence too that hope springs eternal, for we were determined to build there.

We never doubted that we needed new buildings to house our cattle in winter along with the large amounts of hay and straw for their food and bedding. In our minds it was not a question of "if" but of "how" and "when". Even before either we or our heifers arrived at Bank House we had decided what sort of buildings we wanted and that we wanted them in Black Carr. The immediate question was how soon we could get them up and running. With the experience of our first winter very much in mind we were naturally determined to have the benefit of significant shelter before winter could strike again, but there were many obstacles between our clear intentions and their achievement. Only some of the decisions and operations lay in our own hands. We had of course budgeted for building when buying the farm but we still had to deal with the planning authorities, building regulations, the red tape surrounding even the generous subsidies of the day and competition for the services of the building firm we had chosen to employ. We knew of course, that whatever plans we made were at the mercy of the weather.

At the time when we bought Bank House a revolution in agricultural building was well under way. Walls of warm-coloured local sandstone were being superseded by those of concrete blocks surmounted by wooden "Yorkshire boarding" – three-inch slats with half inch gaps between. Roofs of red pantiles gave way to sheets of glaring white asbestos and flagstone flooring to ever more expansive areas of concrete. Furthermore the new process of tanalising wood guaranteed length of life without the use of expensive oak for beams, pillars and rafters. All these changes made for quicker and cheaper construction besides providing much larger, under-cover spaces needed for farming in an age of tractors. Naturally for the sake of speed and economy we were bound to take advantage of the new methods and materials though we departed from common practice in certain important respects most of which related to our youthful, amateur workforce or to our organic principles.

# BLACK CARR BUILDINGS

In brief, our strategy was to commission a firm of builders to erect the skeleton wooden framework clad with "Yorkshire boarding", together with complete asbestos roofing, guttering and fall pipes. By not concreting any of the floor of the buildings we had halved their cost and we saved further money by undertaking all the block-walling ourselves as well as the inner partitions, the water supply and the provision of access for heavy machinery and livestock all round the buildings. The hay racks we would purchase ready-made along with a large supply of loose beams and brackets with which to create our flexible system of pens as varying circumstances required. Altogether these hay barns and fold-yards with the connecting implement shed were to cover rather more than six thousand square feet which was quite generous for such a small farm. They were to be lofty, light and airy for the health of man and beast, and flexible to allow for such adaptation as the inevitable lessons of experience would suggest, a precaution the years amply justified. A key element of this flexibility was to be contributed by the earth floor so that temporary small pens suitable for calf creeps or calving could be put up easily at short notice simply by banging in a few posts and lashing on hurdles or gates.

One might not expect the building of a couple of hay barns and fold-yards to be a particularly dramatic affair. For our contractors it was doubtless just another job, but for us it proved one of the most exciting and exacting aspects of our Bank House venture. The slow, drawn-out drama grew from the tension between alternating periods of sudden progress and inactivity on the part of our builders against a background of our ever more urgent need to be ready for our second winter. As early as our first October we measured the only patch of firm land of Black Carr and pushed in pegs among the tussocks and rushes to indicate the possible place and shape of the buildings. The builders came fifty miles to inspect the site, approved it and immediately phoned the Ministry to report as much. They came again and asked for maps and detailed plans. We spent precious evenings poring over sheets of figures and began to make sketches. We returned to the site and moved the pegs ten feet eastwards. Came November, and with it Ministry approval. At last the grants vital to our going ahead with the scheme were assured. We redoubled our drainage efforts when there was time to spare: found and unblocked a long, six-inch drain and left it disgorging torrents of water from Black Carr's hidden drainage network. December. We carried out pragmatic trials, haying the herd on specific, measured ground to learn how much space was needed for horned and polled cows living together. We had a scary hundred mile drive in sudden snow to consult the builders at their base, seeing all the range of accessory

# BLACK CARR BUILDINGS

fittings they had available and came away with lots of new ideas. *"Horrible drive back with fog and darkness as well as slush,"* says the diary. At Christmas once our goose plucking ordeal was over we had time to mull over the builder's estimate which had just come. Unlike most of such documents it actually added to our seasonal cheer.

After two months of exciting momentum, our building project entered the new year in a state of hibernation from which it didn't revive until the end of April. Nothing seemed likely to happen though with the lambing and calving season we were experiencing we might well not have noticed it anyway. The only timetable agreed was that we should have a barn ready for hay timing, whenever that might be! One day the builders suddenly turned up, left a digger and a concrete mixer and departed. "Perhaps they mean business." we said. Three working days later they had erected ten, twenty-foot wooden pillars in two parallel rows, set in concrete, reaching skyward. We were fascinated to watch them do it by simple leverage without the aid of any expensive machine such as a crane. *"7$^{th}$ May. Barn finished! Looks very fine from all angles."* There it stood, just a roof on stilts but it was a promise of a dream to come true.

Now it was our turn. Down to earth. After levelling the ground with picks, shovels and barrows we spent days filling the end bay of the new barn with old hay ferried down the rough track from the Dutch barn or brought up from the remains of the stack in Square field that had only just survived the December gales. Then we were side-tracked by other demands on our attention principally the annual cycle of shepherding duties stretching from clipping and dipping to sheep sales and tupping besides managing somehow to squeeze in the whole of our first independent haymaking season. A hiatus in the builders' operations ensued rather resembling Parliament's long summer recess but this gap was easier to endure than the last because we now had somewhere to store all our hay under cover and conveniently near to the hay fields.

The builders finally reappeared on August 13$^{th}$ as if they had been waiting for the grouse season to open on Glaisdale moor. This time they really got up steam and in a remarkable burst of activity completed their assignment in seven weeks apart from the making and fitting of hayracks. At this stage the appearance of the buildings was somewhat comical. Only the bottom portion of each of thirty pillars was visible: the rest disappeared behind the canopy of boarding and roofing which they held up. From every angle the onlooker saw right through the buildings and out the other side except where our growing stacks of hay provided some sideways protection from the elements. There was nothing to stop livestock wandering everywhere if they got into the field and nothing as yet to prevent them helping themselves to ad lib hay either inside or outside the buildings. We had indeed got our work cut out if we were to be ready for winter. But we did have approximately three months in which to do it and we were encouraged by recent experience to be optimistic in assessing how much we might achieve in ninety days.

As it happened, for much of that time we had only one assistant so very early on we realised that the herd would have to be in residence before we could even make a start on the concrete block-walling and possibly even longer before we could lay on a reliable internal water supply. Once again we were involved in a race against time, or rather against the weather, but now it was almost entirely up to us.

It may surprise some readers to learn that when we had so much else to do we gave high priority to harvesting rushes and bracken. Many a day we could be seen on the bracken slopes of the farm, scything, sickling, raking and turning before bringing load after load to Black Carr where we were slowly amassing a store of winter bedding. As usual our unconventional ways were influenced by our need to economise. The obvious alternative to our free but hard-won supply of bracken and rushes was to be bought-in straw, and plenty of it! In subsequent years it was to account for a large proportion of our annual expenditure on cattle, but in 1975 we could not afford this having spent the great bulk of our capital already paying for the farm, our foundation stock and the new buildings. As usual we also had other reasons which didn't apply equally to neighbouring farms. One was that removing the cover of dead bracken from what had been good pasture was an essential step towards the revival of neglected land and by using it as bedding we were killing two birds with one stone. Another reason was that this protracted labour-intensive work was appropriate for all degrees of skill and stamina and for however long or short a time. Everyone could help and it was often very sociable.

Rush gathering was a similarly dual-purpose activity. Whereas bracken had invaded mostly on slopes inaccessible to tractors, rushes flourished on lower, level ground where we were applying for substantial grants to pay contractors to carry out extensive drainage improvements. Many of the thickest patches of rushes grew in Black Carr itself and their removal was to prove an extremely helpful preparation for drainage work. The clearance of both rushes and bracken would stimulate what feeble grass survived amongst them and make it more accessible to grazing sheep and cattle.

Harvesting bracken

Once inside the building their roles were complimentary. We spread out a thick mat of rushes directly on the earth floor as a tough protection against poaching from the hooves of heavy cattle. The bracken added on top was to be both warm and absorbent. Though we just managed to accumulate enough of both to be able to bring the cattle in when it was becoming urgent to do so, the work of replenishing the supply of bracken was something we could, and did, continue on and off throughout the winter. Our system was far from foolproof but where skirmishing cows rucked up the bedding from time to time, exposing bare earth, the remedy was simple enough; more rushes and bracken.

For the first winter in the new buildings only about half of the bedding consisted of rushes and bracken: the rest was Bank House straw. On our very first visit in bleak November back in 1972 our predecessor had led us proudly to the door of the intended byre he had concreted for the cows that never materialised. Inside was a great heap of barley grain, a splendid sight indeed, the harvest of a crop contractors had ploughed, sown and combined for him. Of course by the time we bought the farm the grain had been sold but the large, open cartshed was still stacked with its straw for our use. In each of 1973 and 1974 we grew two fields of barley, providing more valuable straw. That from the former year had occupied us in our very first month as we built and thatched a small stack which we sited on the edge of Black Carr, a very solid portent for the barns that we hoped would be nearby. The straw from the 1974 crop we baled ourselves and it gave us a pleasant sense of progress to be able to lead it straight into the gleaming new barn to be stored exactly where it would be used without repeated handling. To the rushes, bracken and three vintages of straw was added quite an amount of spoiled, unappetising hay; some from the stack in Square field damaged by December gales and more from the remains of the Dutch barn supply that had let us down disastrously the previous winter. It looked a lot all together but in the event it proved barely adequate. What helped us to get through was the fact that the herd spent so much time out of doors, and the reason for that was chiefly water:

> *Water, water, everywhere,*
> *And all the boards did shrink;*
> *Water, water, everywhere,*
> *Nor any drop to drink.*

These lines from Coleridge's "Ancient Mariner" might have been written for Black Carr that autumn. There was water everywhere outside where we didn't want it but none as yet inside where we did. Of the two problems the first was the more urgent.

# BLACK CARR BUILDINGS

As the builders left them the buildings were dry inside but all the rain falling on those six thousand square feet of asbestos roofing came pouring down the fall-pipes to be discharged straight onto the ground. What was to happen to it next was our problem. The solution involved us in learning how to install a trap gulley set in concrete under each fall-pipe. These had to be connected to what we called "Offa's Dyke", the long, open gutter that ran down the side of Black Carr and all the way to the beck. The hardest part of this work was digging all the necessary trenches, frequently obstructed by boulders, some of which could only be dealt with by resorting to cold chisels, wedges and sledge hammers. It was such slow going that we abandoned the idea of tackling the drinking water problem before bringing in the cattle, a decision precipitated by the first snow of winter that fell a fortnight before Christmas.

So it was that our long-anticipated celebration to inaugurate the occupation of Black Carr Buildings turned out a rushed affair, its date chosen by the weather rather than by us. That December afternoon while we were hurriedly spreading out the last of the thick bedding all over the floor and putting hay in all the racks for the first time, the herd had taken itself off right up to the topmost fields, out of sight but not of sound. Their distant complaint that we had forgotten to hay them drifted down to us and spurred us on. At last all was ready and as we set off up the bank to fetch them down we couldn't help gloating over the pleasure in store for them. Last night they had lain on the cold, cold field like the raggle taggle gipsies and tonight they would sleep in what was for them the equivalent of a goose-feathered bed.

Our summons from below merely increased their noise but when we continued calling they broke into a stampede straight towards the brink. Could they possibly stop? Or were they fated to emulate the Gadarene swine? They braked abruptly with remarkable efficiency for such heavy beasts and stared down at us in a mixture of disbelief and disappointment. No hay! We were empty-handed. We turned down again, calling all the time, to lead them to Black Carr some hundred feet below and to those strange constructions they had never been allowed to investigate. The way led across a culvert, the last hurdle. Again they stopped. Again we called, but still no hay! Could we be serious? Our diary for the day, usually matter-of-fact, allowed a little embroidery on this historical occasion:

> "Finally Plum believed us, first across, first into the new buildings, cattle all excited and restless, in and out, round and round."

Our excitement was barely less than theirs though less demonstrative: nor was it disinterested. Yesterday, indeed all our winter yesterdays, we had humped hay-bales round the farm. Tomorrow and all our winter tomorrows we should merely throw them into racks close at hand. It was the end of a chapter, and a landmark in our Bank House story, but by no means the end of the job of creating the fold-yards of our vision.

The unfinished business we carried over into 1975 was considerable: some of it demanded attention at once, especially the matter of providing drinking water. It was to be over a year before we could achieve the long-term solution of piping safe drinking water into each pen or division of the fold-yards, equipped with water bowls. In the meantime, the herds had to be turned out every morning and evening and driven across three fields to a tumbling stream of sweetest water. On the way they had to be deterred from lapping stagnant water that lay in Black Carr field, for fear of imbibing liver fluke. This routine was comparatively easy except when foul weather made the animals doubt if they were so thirsty after all. When it came to calving there was no alternative to fetching heavy buckets of water from Offa's Dyke and clambering into pens with them, often in the dark. However, calving occurred only occasionally and when it did we wanted to spend time with the cows and the calves concerned. There was moreover a silver lining to the twice daily chore. By emptying the fold-yards of cattle we had a much easier job distributing the fresh litter and putting the hay ration in the racks before letting them back in towards dusk.

We came across an ingenious and fairly cheap invention which helped us to get water into one part of the fold-yards quite soon: that first January in fact. It consisted of a simple pump in a shallow trough connected to a pipe with its other end in a source of water. For a cow to drink the water in the trough it had to push the lever out of its way, producing about a pint of water thus replacing the water it had just drunk, then a spring pushed it back again, so the manoeuvre was repeated. All we needed besides the pump itself was eighty-two feet of polythene pipe which just reached the pool we created by making a little dam across Offa's dyke. Its success was recorded in the diary:

> "February 1$^{st}$: built dam." "February 4$^{th}$: set up pump." "February 5$^{th}$: watering device seems to be working. Damson was first to be observed to pump deliberately." February 6$^{th}$. Most if not all the cattle use the pump now."

It proved very useful though slow; naturally cattle preferred the stream until the following winter (1975-6) when we built our own permanent system with tanks and water bowls.

## BLACK CARR BUILDINGS

That first season in our new fold-yards was one of ongoing experiment and learning, testing out the system we had devised, finding how best to divide the space for needs of different animals, surviving by trial and error various difficulties arising from our unfinished accommodation. Most energy and ingenuity went on keeping the stock out of the food and bedding stored in the barns close alongside their yards. Hardly a day went by without some new breach in the defences creating frequent little emergencies which took us away from other jobs round the farm, preventing, too, any continuous work on the block-walling round the yards and barns themselves. In our optimism we made a start that January, 1975. We ordered supplies of sand and gravel, and, of course, the concrete blocks themselves though the cement we fetched piecemeal from a village builder, distrusting our chances of keeping it dry for long. We got so far as to dig two trenches between three pillars and laid the concrete foundation for the block-walling to come above. Alas, what Prime Minister Macmillan called "events, dear boy, events", overtook us and that turned out to be the end of our first walling season. Those two trenches, having filled with water, had to wait for some nine months for further attention. However, our purchase of hundreds of concrete blocks had not been in vain, even in the short term. Remembering our childhood play with wooden bricks we stacked the blocks, bonded loosely to form a thick, temporary wall right across the west end of the buildings where it proved remarkably effective in keeping out the prevailing wind coming off the moors and funnelling down the dale. It gave us a reassuring foretaste of the haven the stock-yards were to become when completed.

There was no comparably simple solution to protect the hay in the barns while the herd was in occupation, for to build substantial lasting barriers we would have to remove the existing ramshackle patchwork of defences. So the daily battle of wits continued well on into the spring, a battle in which any superiority of wits in our part was easily countered by the unrelenting persistence of the cows, to which was added the awesome strength of our first bull early in the new year. At least we began to work out how we were to protect the hay the following winter. The hay-racks supplied by our building contractors – who didn't deliver the last ones till the December week we brought the herd in! – formed an effective, continuous bulwark between the barns and yards. Our Achilles' heal was the gap below the racks down to the ground. Our plan was to make palisades to cover this gap out of cut-offs bought very cheaply from sawmills. These were to be lashed securely to horizontal beams using plastic or hessian baler twine, the famous "bit of band" that is the panacea of all dales farmers whose pockets rarely fail to deliver.

Palisade construction and concrete-block-walling were alike put on the back burner. Later they were to dominate preparations for the winter of 1975-6, our third at Bank House and our second in Black Carr buildings.

Meanwhile there was an important part of our work towards completing the buildings that we could, and did, pursue straight away: making a rough but serviceable roadway from the main track up to the regular entrance to the buildings and round to the new implement shed. It is surprising how many trailer loads of builders' rubble it takes to make solid foundations for a farm road. Fortunately it was free. Builders are often only too glad to have somewhere to dump unwanted remains of buildings they have demolished so thanks to this example of mutual benefit our road-making resulted in mucking out and hay-timing becoming easier and more reliable in the coming summer season of 1975.

The letter we sent out to friends and relations that Christmas, summarising our experience and progress during 1974, referred to the newly occupied buildings in Black Carr as *"our greatest achievement so far"*. So they were. Other projects may have taken more time and hard work, occupied more people, taught us more lessons or shown us quicker results but getting our fold-yards, barns and implement shed to the stage at which we could begin to use them, even if they were not yet finished, was more important and marked our greatest progress. It represented a heart transplant, a shift in the farm's centre of gravity from its eccentric position on the very edge of our land and on its awkward ledge on the hillside, to a site commanding easy access to all our level, most fertile fields. The major farming operations such as ploughing and cultivating, muck-spreading and haymaking, would all benefit as would our care of livestock when for example bringing in cows from the further pastures to calve or receive medical treatment under shelter. A great deal of internal traffic was now eliminated; notably the movement of heavy equipment, of loads of muck, hay or straw besides sheep and cattle on foot. All these previously used the steep, winding track on Boston Bank which snow and ice frequently made impassable in winter.

Of course it was not all gain: there was a price to pay and it was we who had to make the journey between the farm house and Black Carr morning, noon and night, usually on foot, but it was definitely worth it and never dull for we had a choice of six genuine alternative routes and trudging up from the buildings for breakfast and lunch could be given extra purpose by looking for sorrel to add to the soup, blackberries for our supper or firewood for the evening fire. Each route gave slightly different views of the complex of buildings below, all of them attractive since we had taken great care to

## BLACK CARR BUILDINGS

design them as a group of related shapes and angles. It wasn't until all the block-walling had been completed way on in 1980, that they looked their best at which time we wrote of the block-walling:

> "Visually it is very satisfactory as it makes the whole cluster of buildings seem rooted in the ground, solid and protective and it certainly keeps out a great deal of the weather".

Perhaps the most successful feature of the buildings was the Yorkshire boarding. Anyone not familiar with this system might expect the half inch vertical gaps between the long three-inch wide boards would allow a most unwelcome amount of wind and rain to blow straight in, to the discomfiture of the beasts within. Not so! We have stood inside, only three or four yards from the west end boarding, while a stiff gale was lashing rain against it yet we could hardly feel the air on our faces let alone any of the rain. Apparently the greater volume of wind bouncing back off the boards dragged with it the smaller proportion of the wind and rain that was making for the gaps. Only lightly drifting snow found its way through, besides of course a wonderful amount of sunlight that made the barns and fold-yards so pleasant to live and work in. There was no need for windows as the boarding with its gaps acted like lace curtains. At night moonlight lit up each great recumbent form of the cattle as they lay puffing regularly, chewing the cud on and on. Such a peaceful scene made late night visits to check on possible calving a joy as much as a chore.

All in all, as our plans on paper gradually materialised in the form of walls, roofs and additional fittings which we could begin to use, we were gratified by the extent to which we had got it right, right that is for our unconventional needs and tastes, though they wouldn't have suited many other farms or farmers. Most importantly we never seriously questioned the wisdom of siting the buildings in Black Carr. Surely the proof of the pudding lay in the eating and whatever the bias of our personal verdict that of the cattle was definitely thumbs up, and the proof of that came with successive calving seasons, starting in 1975 when we didn't lose a single cow or calf. This startling contrast with the previous season was partly a measure of the early improvement of the farm as a whole but much of the credit was due to the new buildings in particular and to the relaxed way the cattle made themselves at home there.

Black Carr Buildings

## BLACK CARR BUILDINGS

It seems appropriate therefore to end our account of the creation of Black Carr buildings with glimpses of life in the fold-yards at calving time. To limit the distorting effect of rose-tinted spectacles that so easily colour accounts of time past we shall take these last pages from the farm diary and a letter written not long after:

*Jan 11*[th]   G found Heather with a calf staggering about in main fold-yard. Fetched M and J. Constructed pen by fetching hurdles from Holey field.

*Jan 12th*   Brambling, Heather's day old calf much stronger. Knows how to suck. Can jump and skip already. Beautiful chap.

*Feb 11th*   Duchess produced her second heifer in middle of Black Carr rushes, unaided and silently so hardly noticed by nearby workers. G carried calf into pen in buildings.

*Feb 12*[th]   Duchess' new daughter very fit, filled out, sucking and skipping – will soon want to be out of the pen.

*March 17*[th]   G made calf pen at west end of BCB and put Damson in – beginning calving. D's calf began and stopped. Dusk. G & M with lantern and cords, great drama getting calf out – very big heifer – alive and kicking – in Rembrandt gloaming and lantern light – pulled by J & G assisted by M.

*April 16*[th]   Helpful and Edna – at her request – put in Mushroom (field). Edna calved a heifer unaided during lunch. Left out there overnight. Afterbirth OK.

We hadn't time to write long letters in 1975 so for a more detailed picture we end by quoting from a later year but conditions were very similar and what is described can stand for any period in Black Carr Buildings' history:

"*February 27*[th] *- For the moment it is all calves. Each of the last three mornings I have gone down to BCB to feed the cattle and found a newborn calf. Edna on Friday, Persuasion on Saturday, Damson this morning. I do wish you could see them at this stage. They are at their tamest when very young, they haven't learnt to be shy and fearful so you can stroke their beautiful silky heads; under the throat or on top where the horns will come is the most appreciated. We now have the three senior Christmastide calves beginning to eat hay seriously and solemnly – and then there are seven February creatures; Dinah's heifer with a thick curly coat, black burnished with red; Duchess's pale, tall and reticent; Harmony's, also a son, sturdy*

BLACK CARR BUILDINGS

*and adventurous leading the races from end to end of the barn, and glad to suck your finger; Dolly's daughter, nearly always curled up, asleep, under a hay rack, half-covered with hay and straw. These four have been at large for four days while the newcomers are still in pens.*

*Meanwhile Blanche is either a fortnight overdue or a week early, looking vast so you can imagine twins; and then there's Asphodel, first time calving, due today but not quite ready, with no bag at all, she's the one most likely to prove awkward and need help:*

*We shall make an extra pen tomorrow. Then it's March on Tuesday and Tulip is due on the 3<sup>rd</sup> and Flora on the 7<sup>th</sup> and we shall be bringing her up to the yard to augment our milk supplies since we have a larger population than usual.*

*And in the midst of one set of calvings the next is being prepared as Frolic has three heifers as well as Jeannie to cope with; he was romantically close to Jeannie all today – very polite and proper as far as we could see but she's a discreet lady so all we have to go on in determining next Christmas's calving date may be that in three weeks' time Frolic and Jeannie don't want to know each other."*

With the help of Black Carr buildings we had redeemed our calving record from its tragic beginnings.

## Chapter Twelve
## JEANNIE

Friday 11<sup>th</sup> January 1974 should be regarded by a historian of our Bank House venture as a red letter day marking a most important step towards establishing a pattern of living and farming we had set ourselves. Yet the same historian could be forgiven if skimming through the diary for that day he underrates its chief significance: *"M tidied house – negotiated transport for Jeannie and got her installed and D and M got her milked (6 ½ lbs)."* The next day's entry was more helpful: *"Jeannie provided cream and milk – no more fetching milk from across the dale!"*

The range of milk delivery service in Glaisdale extended only a mile up the dale from the village. The best arrangement we had been able to make had been for our milk to be left outside someone else's house three times a week, still a mile away from Bank House. For our first five months every occasion taking anyone off the farm had to be planned with the possible collection and carriage of our milk in mind. This inconvenient and unreliable system was sufficient reason by itself for us to celebrate the arrival of a cow already in milk and well used to being milked. Naturally our barely tame herd of suckler heifers had not been able to help us out yet in this respect, though subsequent shortages due to drought or surges of visitors were briefly eased from time to time by raiding the suckler herd. Meanwhile we had to remember which day of the week it was and go and fetch the milk, usually at the end of a tiring day, hence the note of triumph in the diary.

During ten years of married life at Abbotsholme Giles had come to recognise that for Mary it was an article of faith that proper milk came from farm buckets or lidded cans not from factories and glass bottles. Nevertheless he had been surprised in September 1972 to find her snapping up a bargain piece of simple dairy equipment while on a break from school in Reeth in Swaledale. After all we owned neither land nor cow and all our searches for the former had so far proved fruitless. It became clear that wherever we should go after Abbotsholme, whatever we should end up doing and with whomsoever, Mary was determined we should have at least one cow to milk for which we should need cottage scale dairying equipment. True the cow was missing, but at least we now had something to pour her milk into.

The name of Jeannie may ring a bell in the attentive reader's memory as that of the house cow who came as part of a package deal with Botton, along with a probation lad, Dick Bishop: she in mid-lactation ready to milk, he hastily trained to do the milking. Of all the cattle we were to keep in our eighteen years at Bank House Jeannie was the most unforgettable, contributed the most to our survival and left her mark in more senses than one. Long after her demise the byre where she was always housed along with the splendid succession of her calves is still known as "Jeannie's place". We got her cheap because she was due for slaughter, not on account of her age (she was four years old and had had only two calves), but because she had lost a quarter of her udder from mastitis and had a weeping, gammy eye – we were never quite sure how much she could see out of it. However, three quarters of a dairy cow's yield of milk for a single household can be a bargain at a bargain price, which in this case was fixed as the contents of a small, damaged haystack of ours, some of it distinctly damp. Botton collected it at the end of February by which time Jeannie had become a familiar and key part of our community. Incidentally she was to lose a second quarter to mastitis two years later but it never caused us to revise our opinion of our bargain.

On the rare occasions when Jeannie mingled with our Devon heifers she cut a very distinctive figure, not just because of her typically shorthorn roan coat – brindled silvery white with reddish brown – but in her form and behaviour too. Tall, bony, gaunt and graceless, without much in the way of plump flesh to gladden the heart of farmer or butcher, she would stare uncompromisingly through her good eye as if to assert that she didn't care for the friendship of other cows or us people, only for her calves. Her powerful maternal instinct was her only endearing feature but in this she was magnificent. It was in trying to kick another cow's inquisitive calf away from her udder that she caught Mary's knee instead while Mary was milking and anchored to her three-legged stool. Thus befell the most serious and painful accident of our years at Bank House, incapacitating Mary for five weeks in plaster, and bequeathing her a weak knee thereafter: but that was another story.

Nevertheless our abiding memories of Jeannie are of pleasure, satisfaction and achievement, for milking a cow by hand is not just a useful job but an intimate, sensuous experience too, which only becomes drudgery if prolonged by great numbers and overlong hours. With only one dairy cow to milk we had the satisfaction of slotting another central piece into the hub of our farming jigsaw puzzle.

Jeannie and son: "Don't you dare wean him"

# JEANNIE

Besides fulfilling another of our ambitions, milking with its attendant activities and by-products added another dimension to the varied, active life we were able to offer our younger visitors and long term helpers. For Giles it was something new and exciting but even Mary recalls that if ever down-hearted she would console herself by milking Jeannie and pouring the booty into the bowls waiting on the dairy shelf. Jeannie's arrival changed life at Bank House in many ways. In the first place it imposed an unfamiliar discipline on our daily routine. Since she came without a calf we had to milk her morning and evening, and late mornings had an unwelcome knack of producing late nights. We had to fit these two sessions into our long, spontaneously erratic working timetable. It also imposed an element of hygiene for which some of us were not renowned, though this was a small price to pay for the rewards. To have an abundance of fresh milk for our own use was riches indeed and contributed hugely to the health and growth of man and beast. Though we had always reckoned we fed ourselves well, the boon of Jeannie's creamy milk helped to satisfy our considerable appetites even more effectively in many ways. We were less conscious of the cost when sloshing milk onto our breakfast porridge or cornflakes; quiches appeared regularly on the menu as did chocolate custard and junket, while fruit dessert was often embellished with cream.

The taste of milk varies distinctly from breed to breed, even from cow to cow as connoisseurs assure us, especially when the connoisseur is a calf. Mary had been on tenterhooks when the first bucket of Jeannie's milk was brought into the dairy and stood to cool. What would it taste like? What an anticlimax if we didn't like it! She needn't have worried. Everyone pronounced it superior to the bottled milk it superseded and fully justified our choice of the traditional local breed, the Dairy Shorthorn.

The cats were first and foremost among four-legged beneficiaries, followed closely by the sheepdog Kim. Whether his evening pan of flaked maize was soaked in milk or water – and he never seem to notice which – depended on the balance between the comparatively steady yield of milk and our fluctuating animal and human demands. For instance, priorities were liable to be turned upside down in lambing time whenever there were pet lambs to be bottle-fed. On the other hand there were times when supply out-stripped our normal consumption. It might be that a new-born calf had over-indulged, temporarily lost its appetite and needed to fast for a day or so. Or maybe the rush of spring grass doubled the amount of milk just when our household was at its lowest with people away. No problem! at least not after the end of that January when Jeannie came, for she was soon followed by our first breeding sows in the rotund forms of Matilda and Placidia.

# JEANNIE

Thenceforth there could be no waste, for everything that had been or could be food was consumed with noisy relish, while we onlookers had the additional pleasure of watching scraps and surplus being converted into delicious bacon and fertile compost.

Unfortunately for them the pigs came well down the list of alternative destinations for our milk. Our everyday routine involved leaving the fresh warm milk to cool in the dairy in shallow bowls while the cream rose and then skimming off the top layer into little jugs and pouring the rest into quart and pint pots for general use. When it was clear we had a surplus the easiest way for us to deal with it was to let it stand until it began to curdle, and then hang it up in a fine, strong cloth looking like a Christmas pudding. Once it stopped dripping there was little else for us to do than add a little salt before enjoying eating the cheese, hoping the while it would fall to someone else to wash out the slimy cloth. This simple "cottage cheese" didn't keep all that long but then it never had to! The whey that dripped through the cloth ever more slowly into the basin beneath was the nearest the pigs ever came to enjoying a milk diet but it was very nourishing nevertheless.

Time was, before the coming of railways and lorries made farm to farm collection of milk a practical proposition, when making and selling butter provided a major part of the incomes of most dales farms. By the time we arrived on the scene butter-making was a rarity, almost a hobby, and home-made butter a treat. We were able to catch the tail end of the tradition on the very smallest scale. On occasions of particular surplus, especially in early summer, we were sometimes tempted to make our own butter but only if we had time to spare for it could be a tedious and uncertain business. Never a money spinner, it appealed to us and most of our helpers as an experience. Having left the milk for as much cream to rise as was possible we skimmed it off into a square, glass mini-churn held on our knees or on a table. It needed strength of grip and great patience. If turned in arrogant expectation or even frustration it would never "make" however long the unsympathetic churner churned. Part of the fascination of the process was its element of magic. The moment one decided it wouldn't work, low and behold! a giant lump of butter swimming in pallid buttermilk. Thereafter it was straightforward, slapping it into the desired shape and patterns with wooden pats and setting aside the nutritious buttermilk to make into scones. Our home-made butter tasted as different from its standard factory counterpart as Stilton or Gorgonzola does from plain cheddar, causing some faces to beam with pleasure and others to wrinkle in suspicion.

To return to Jeannie. We are in danger of making it all sound too easy. Just buy a suitable cow and you have your own milk supply for the rest of her productive life. It was never like that. We knew only too well what a vast amount there is to learn about the proper care of a house cow, some of which Mary knew already, much we learned from good neighbours but even more from Jeannie herself. One very necessary skill we had to acquire was that of securing a continuous succession of calves without which the milk supply would dry up. This involved managing the "dry" periods so the cow could enjoy her necessary rest between calves while avoiding too long an unproductive interval. Not to get a cow served at the right moment can be very expensive and this was particularly difficult to avoid with our house cows kept up near the farmyard while the bull was occupied with the main herd at the far end of the farm. At one time we were getting worried we might have missed out with Dora, a reticent, new house-cow that seemed to conceal all the usual signs of "bulling". One day Mary watched her standing still for quite a while, staring vacantly towards the distant herd without any expression of her desire or need. Acting on nothing more definite than Mary's hunch we left our other work and went to fetch the bull. As we cajoled the unwilling lord away from his hareem we couldn't help wondering if we were creating extra work for nothing. However there was no room for doubt once he arrived. Our rewards were instant visible proof and a fine calf in due course.

With our house cows, as with our suckler herd, we departed from common practice. Our aim was not primarily to obtain maximum yield and profit, rather to achieve a happy balance and agreement between the interests of the cows with those of our bank balance. It seemed to us that our result was a life of less strain all round and this was borne out by the health and longevity of our stock. Jeannie exemplified this convincingly. In addition to the two calves she had already brought into this world before we saw her at Botton in December 1973 she reared a further ten calves for us and often they were the best grown of their year. Flora's birth was recorded as follows:

*"December 18$^{th}$ 1975... M & G just going to bed: heard a moo. Jeannie had taken herself to the bottom of Boston (bank) and calved unaided, mooing possessively in triumph not pain.*

*Lizzie, midwife, also in the snow.*

*Calf got under a thorn bush. J worried so G carried calf up to J's byre. All happier ! -except G's eye scratched in said thorn bush. Bed 12.30 am."*

# JEANNIE

Flora was a Devon/Shorthorn cross heifer which we kept to breed from as an addition to the main suckler herd, yet she in turn raised twelve calves for us, was an amenable, temporary house cow at times and was still alive when we left the farm in 1990. So Jeannie and her daughter spanned all our eighteen years at Bank House between them. Furthermore, length of life and quality of offspring were not Jeannie's only assets. Another was the regularity, almost the predictability, of her calving. For seven consecutive years she calved within the period November 27 to December 25$^{th}$ which made it much easier for us to guarantee our milk supply and to leave us a calculable gap to cover with companion house cows.

Dear Jeannie! What a blessing she proved as a settled house cow who knew all the ropes and could be relied on to help us to teach generations of young people how to milk. The legend of Mary's knee merely added spice to their apprenticeship as well as respect for Jeannie without posing real difficulties. Just how fortunate we were in her, the first and cheapest of all our house cows, was emphasised by Lizzie, a second shorthorn from Botton that we bought to cover Jeannie's dry period in September 1975. Immediately she was unloaded into our yard she burst a hurdle and cleared a cattle grid to join our neighbour's herd beyond. She was still thoroughly unsettled the next day which she spent beating the bounds, hardly grazing at all. It was clear within ten seconds of their first encounter that she was to be boss over Jeannie and it was deemed prudent for Giles to be fall guy as her milkman, a job which he had never done before. It didn't seem to matter that he was clumsy and inexperienced; an adopted, totally unjustified voice of authority made up for it. Ex-school masters have their uses!

Lizzie wasn't destined to boss Jeannie for long. She died suddenly and without warning. It was never explained either by our vet or by a post mortem and remained one of those baffling losses that leave one none the wiser. Her place was taken by Dora, another Shorthorn though red, and without horns: a gentle creature that kept Jeannie company for seven years, giving us nine calves in addition to the pair she had borne for her previous owner.

By 1983 we began to wonder how long the old guard would last so we planned for Jeannie to produce her own pure Shorthorn successor by getting her served by AI. Ellen was born to be heiress apparent. As part of her royal education she was halter trained and taken to Egton Show where she won a rosette in an admittedly small field. The thought that Ellen would produce her first calf the following year made it easier for us to decide it was time for

# JEANNIE

Jeannie to call it a day. She was rheumaticky and finding our steep banks increasingly difficult so we took the hardest step livestock farmers have to take, and bid her farewell early in December 1984. Even as scrawny meat she earned us more than she had cost us eleven years before. She wouldn't have wanted us to shed tears but we had lumps in our throats.

The epilogue to this account of our partnership with Jeannie is not what she deserved and far from what we hoped for. The calf was never born. Ellen took ill and died after a fortnight's vain treatment and nursing, during which it was discovered she had a major internal abnormality. Thus the Jeannie dynasty was cut short before it had barely begun, and alas there was no longer even an ageing Jeannie to bear another replacement.

## Chapter Thirteen
## HOMAGE TO BONUS

Neither Mary nor Giles had had any experience with bulls before coming to Bank House. Mary's best preparation had been that of coping with excited stallions: for Giles, less helpfully, it had been as a wing-three-quarter on the rugger field, cutting off escaping opponents from the promised land. Nevertheless we were not deterred. Placidity had been prominent among the advertised established merits of the Devon breed when we chose them for our farm. The common reputation of bulls in general as dangerous, temperamental animals applies more particularly to dairy rather than beef breeds like ours, a fact corroborated by official regulations regarding the presence of bulls at large where there were public footpaths. It seemed obvious to us that the aggression of a bull was greatly increased by his being kept away from cows or, worse still, shut up alone in a building where their sounds and scents could still reach him, which in fact was almost inevitable. We determined that any bull we should have would live with female company as much as possible, indoors or out, even if that company was occasionally reduced to one. Of course caution would have to be the order of the day for everyone near our bulls but in the event it turned out to be caution untarnished by fear and resulted in mutual respect and relaxation between human and bovine once new bulls had settled in.

Not every dales farm breeding cattle had its own bull. The service of artificial insemination provided by the Milk Marketing Board was efficient and comparatively cheap. As long as the period of a cow's bulling was noticed in time a farmer had only to phone by breakfast for the AI man to appear that day and there was a strong likelihood pregnancy would ensue. With a dairy herd coming in to be milked twice daily it was not much extra trouble to have the bulling animal ready penned. In contrast, with beef herds such as ours, it would often involve extra treks driving an excitable cow from the further parts of the farm and back again. We realised there would be distinct advantages if we were to have a bull in permanent residence with the rest of the herd. He would be far better than us at detecting the right moment and far less likely to be distracted from his business than we were bound to be at times such as haymaking and harvest. It appealed to us that keeping one's own bull was a more natural way of maintaining a herd than AI and would fit in with our whole approach to farming. In any case, we could always fall back on the AI service when necessary; a policy of belt and braces.

Ever since we bought the farm we had considered these matters without taking an irrevocable decision but lived comfortably – at a distance – with the idea that managing a bull would be one of the many challenging aspects of our adventure. "At a distance" because we didn't have to do anything about it straight away while so many other things were more urgent. All the heifers making up our original herd should have been in-calf before coming to Bank House, so that we were not expecting to need a bull until they had produced their first calves in the Spring of 1974. Even so, we began making enquiries with a view to buying a bull soon after our own arrival, thinking it prudent to get him settled in well before he was due for active service.

By far the biggest sales of Devon bulls took place in Exeter but we didn't think we could spare the time and energy to go so far during those first hectic months. Then we learned that there was to be an alternative sale at Perth on October 25th, so after a full day's work and a couple of hours' sleep we drove off to Darlington, caught a chain of night trains, attended the sale for six hours and got back home at 9.30 pm with the roar of bulls and the hubbub of the auction still ringing in our ears. Oddly enough we were thankful that our single, tentative bid had not been successful. We realised we were not ready. Nevertheless it had been a valuable trip in terms of what we learned, and we met the secretary of the Devon Cattle Breeders' Society who had helped Mary to buy our heifers and realised that in future we should do better to negotiate through him and avoid the hazards of going to bid ourselves.

So it came about that our first bull arrived at Bank House on the last day of February without our ever having seen him before, indeed we hardly saw him then for it was long after dark and six inches of snow was just beginning to fall. Two days later Giles wrote to his parents at length to put them in the picture, knowing they would not be satisfied with the brief entry in the diary he was in the habit of sending them.

*"We are now the proud and trepid possessors of a bull. A phone call on Thursday morning warned us that our Mr Pascoe had bought us Barton Bonus, lot number 34 at the Exeter sale and that he was already in transit. It got dark and no van, no bull. We imagined him somewhere on a motorway and expected a phone call asking for instructions how to get here. But in the end the headlights came up drive and an enormous vehicle appeared in the dark, and the wrong bull was let loose down the ramp, a gigantic creature with number 8 from the Prince of Wales' Farm on his posterior. Luckily for us a neighbour had found this van stuck before a railway bridge that was two inches too low, so he escorted the van round a four mile detour and arrived with it to help unload a bewildered beast.*

# HOMAGE TO BONUS

*We left him to recover in the main part of the stable with the curious neighbouring noises of geese and pigs on either hand and he seemed content to do nothing all Friday. Today we took our courage in both hands and ventured in with a halter and a bowl of barley. He was nervous and took a lot of encouraging and coaxing before allowing me to rub his shoulder. We got the halter on, led him out and up to the trough, but he was not calm. The running water, the ice of crushed and refrozen snow underfoot and the fall of snow melting off the roof made him want the comparative familiarity of his stable retreat, so we re-entered in haste. I can't pull him at all, only act as a rather weak brake and deflector. We had a second slightly easier repeat performance this afternoon. Well, we've made mutual acquaintance and I hope it will be more relaxed a relationship very soon."*

Our relationship with Bonus certainly got more relaxed, but little else did, in fact it is hard to imagine circumstances less conducive to the calm development of confidence between a potentially excitable beast and his novice masters, than Bank House Farm that March. The general background was our first exhausting lambing and calving season, already described, with its hectic drama of life and death going on all around him day and night. Even closer at hand were the raucous and shrill noises of the pigs and geese sharing his stable which rose to a deafening pitch whenever there was expectation of food or of being let out. Less worrying but just as puzzling to him must have been the great number and variety of human voices, people he didn't know and couldn't see; the voices of our household, our visitors, our neighbours, our vets and then, our builders. Yes, builders! After we had waited for them through the quiet months of winter when any bull would have heard little to excite him, our builders had chosen the week of Bonus' arrival to move in with their cars, lorries and supply firms to embark on a major programme which included taking the roof off the farmhouse. They turned our yard into a Piccadilly Circus, confusing traffic with livestock for over a month.

It was against this background that we set about training Bonus to adapt himself to Bank House and to adapt ourselves to him, relying principally on instinct and common sense, but restricted to the snatches of time between other jobs. Naturally it was a gradual process. On his third day, a Sunday, we tried to get a halter over his head but he declined repeatedly and skilfully yet politely. Then a classic opposition of male testosterone shattered the peace. Monsieur Mollet our gander, co-tenant of the stable, thought his honour was impugned by the presence of this gigantic newcomer so near to his ladies and showing a fine disregard for his own safety, and for the odds against him, flew into the attack. Bonus reacted in a puzzled sort of way and though the result was not the knockout blow it could so easily have been it was a decisive victory on points for Goliath. The valiant David had to be removed from the ring with a bloody foot which never fully recovered.

Since the pigs and geese as well as Bonus all needed to be exercised in the yard we had to juggle the timetable. With the combination of bull and geese now ruled out we tried letting the pigs out with the bull; the diary records that the pigs started to get knocked about. So instead we put the pigs on the Knoll with the main herd of calving heifers, but the pigs got into the garth and upset the ewes with lambs, which put paid to that. We were trying to tame Bonus to the point where it would be reasonable for us to introduce him to the main herd up on the open bank side. The eagerly awaited day was judged to have arrived on March 22$^{nd}$ and at last our plan of having our bull living with his hareem all the time was established. This change was made possible by the recent completion of the long new fence, cutting off the herd on the near bank from the nutwood and all the fields beyond.

There followed nearly two months of what retrospectively may be regarded as a sort of honeymoon period when our duties as bull owners seemed comparatively plain sailing. Bonus lived happily as newly-weds are supposed to do. He was properly occupied with his wives and whenever we needed to move them on to pastures new he plodded along contentedly with them. Butter wouldn't have melted in his mouth. On our part we fell readily into the new regime in which our chief daily responsibility was to witness and record how, where and when he bestowed his favours, so that in due course we should know when to expect each cow to calve. This proved more complicated than we expected. One day we would find Bonus in a mêlée, besieged by three importunate ladies at once, for a cow in season rouses not only bulls but other cows as well. On such occasions it took close scrutiny to be sure which was the real prima donna of the day. At another time we would find Bonus grazing closely alongside the same cow day after day without giving us a clue as to whether we should write down all days or none. To resolve the complication that we needed to record the degree of Bonus' interest and activity not just the date, we invented a system of notation borrowed from music. It ranged from *piano* (*p* or *pp*) for mere proximity or slight interest through to the grand, passionate and prolonged carry-on we designated as "*f*" or even "*fff*" (*fortissimo!*). The system worked very satisfactorily being easily understood and employed by all concerned. Our observations were written down in a grubby pocket notebook under the heading "Bonus' Amours" and the first succinct entry was, very appropriately – "March 30 Jeannie *f*."

## HOMAGE TO BONUS

The end of the honeymoon period was heralded by two innocent remarks in the diary for May 27$^{th}$. "*Bonus got across to Wood's field to attend to a lady.*" "*Giles got him back.*" One of our visitors said she heard two cows calling him. What Bonus had *"got across"* was the winding beck which for half a mile served as a nominal frontier between farms. The cows calling him belonged to Fred Wood's dairy herd, and *"got him back"* with its suggestion of an easy passage was adequate for that occasion but certainly wouldn't be for the long sequence of retrievals and expulsions that were to follow. Bonus' first excursion was to prove the opening shot in a border war to be contested on and off throughout each summer for three years or more, at times testing our perseverance, courage and wits to the hilt.

These protracted campaigns resembled the traditional marauding raids immortalised in the Scottish border ballads in several ways. There was raid and counter raid and in each case the effective spur was the lure of ill-gotten food supplies and fair ladies. As with the historical precedent there were periods of intensive lawlessness repeated almost daily, interspersed with others of uncanny calm and peace though nothing had been solved or forgotten. It was also a war in which there was a bond of fellowship and respect between the raiders on either side. The real foe was us farmers.

Except after heavy rain the beck was shallow and stony with rippling water running between low, tree-lined banks, which it repeatedly washed away in places. As a frontier it was respected by most sheep, though Desdemona took her small lambs across with her as she pleased. With cattle, only the contented and law abiding were deterred unless there was effective fencing as well which for the most part was lacking.

As was to be expected Bonus' appetite for adventure had been whetted by his first crossing of the beck. The diary for the following day reads:

"*G got Bonus out of Wood's fields twice – but left him there the third time, having watched him get through the fence immediately after it was repaired. G saw Mr W.    Not very bothered.*"

How fortunate we were in our neighbours!

A temporary remedy after a third consecutive day of truancy was to move Bonus and our herd to a point further away, from which distance he continued calling without breaking bounds. As with so much else at that time we were on what nowadays is called a steep learning curve as to our bull's nature and character. We saw no hint of temper, far less of rage. He simply moved his great bulk through our petty obstructions with majestic power to where he wanted to be.

We were entering a new phase of our border war, one which required a new strategy. A Maginot Line of impregnable fencing was out of the question, so instead we pinned our hopes on electric fencing of which we had no experience at all and had to learn everything from scratch, mostly by trial and error. The theory is simple. Animals dislike electric shocks. Buy the apparatus, complete with battery, and attach it to a long wire stretched out before the forbidden territory, hung on little insulated stakes. Switch on so every time it ticks the wire becomes live for an instant. Voila! However, erecting or repairing electric fencing is time-consuming, so is the process of teaching animals to avoid it. We remember T.M. chuckling because when he replaced the wire of an electric fence just outside his milking parlour with an ordinary string without any battery, his cows couldn't tell the difference and continued to avoid it from habit. That was the reward of years of training, not something we could copy.

We soon learned the limitations and drawbacks of electric fencing. If one animal pushed through the wire and broke it the whole fence became harmless and all the other animals could walk through it with impunity until it was repaired. Only the first transgressor got a shock and might be deterred in future. Any small branch off a tree connecting the wire to the ground short-circuited the system. If the wire was set too high small lambs could go underneath without touching it; if set too low cattle could step straight over it, especially a bull, and on undulating ground it was difficult to avoid one fault or the other. In any case sheep's wool was protective, and an electric shock which affected inquisitive noses hardly penetrated their fleeces, so when sheep and cattle grazed together the sheep could let the cattle through. We soon learned how to test whether a wire was live or not by touching it with the tip of a piece of grass and gingerly sliding it forward. The pulse of the battery, hardly discernible through the length of the grass, increases as the gap between fingers and wire diminishes until, ouch! And when a battery runs down, usually on a Sunday, the animals are sure to be the first to know it.

For us it was a case of eternal vigilance. Whatever our other work, when it was discovered that the electric fence or a stretch of our barbed wire had been breached someone had to down tools and rush off to restore it. Gradually we taught Bonus that the odds on his getting stung were increasing till at last one day he seemed reluctant to push his luck. We were winning! Or were we? Within two days our easing struggle to keep our bull in was overtaken by a new challenge; to keep another bull out! Fred's bull had found his way across the beck into our fields. From then on it was two-way traffic and of course, if he were near enough, Bonus took full advantage of any breach by the other

## HOMAGE TO BONUS

bull entering Bank House to return the compliment. It must have been a matter of sheer chance that the frequent excursions by the two bulls never coincided to bring about a direct confrontation, not, that is, in Bonus' reign. Lady Luck also saw to it that none of our pedigree cows or heifers ever produced a calf sired by Fred's bull whereas when two of his cows bore unmistakably Devon calves he seemed delighted.

At one point invading cattle put our defences out of action seven times in eight days; shortly after, Bonus was serving a heifer across the beck! Fortunately none of this diminished the harmonious human relations . Neither farm had the resources just then to put up really stock-proof fencing that would have returned the border warfare to its proper place – history. We had agreed not to be at all legalistic over our beck-side difficulties. We all felt free to carry out immediate repairs whether a post or two or a new stretch of barbed wire on one bank, or a reinstated electric fence on the other, whichever was easiest, and also to drive each other's cattle home. There were periods of respite when the beck-side fields were not needed for grazing, or too well eaten down to be attractive, but the only guarantee of lasting peace was brought by winter when cattle needed to be under cover indoors. For this we had to wait for the first December snow when Bonus went into our new foldyards along with his ladies. However we didn't wish to risk keeping a creature of his height and strength in the unfinished Black Carr buildings any longer than necessary. Our precious stacks of hay and straw were protected only by strong hayracks he could almost reach over and by palisading we had hastily knocked together out of off-cuts from a wood yard. As February and our second calving season approached, when none of the cows wanted his attention, we brought him up to the yard to spend the rest of the winter with Jeannie and the house-cow brigade, always in loose boxes at night, sometimes out on the rough banks in daytime as weather permitted.

The date for turning out cattle after being wintered indoors naturally varies considerably from region to region, season to season and from farm to farm. On this north side of the North York Moors a typical year will see most herds emerge to skip and gallop and stretch their legs about the second week in May. By that time in the late spring of 1975 our most vulnerable beck-side fields were already sprouting barley or laid up for hay so Bonus had no early chance to return to the scenes of his watery wanderings. Sad to relate it was his fate to go no more a-roving late or soon. On May 17th he had run a high temperature and stopped eating for a while. With the vet's help he recovered and we were heartened by what is usually a sign of well-being when we saw him mounting cows once more. But on July 15th he took ill again. We brought him back to the stable and found he had a temperature of 104ºF instead of the normal 101. Two days, two vets' visits and several injections later, he died. Giles had taken down a drench left by the vets. Bonus was standing still, under his favourite tree but when Giles had one arm round his head and was trying to push the bottle into his mouth with the other, Bonus suddenly collapsed and was gone. In a letter giving the sad news to his parents Giles wrote:

> "We have lost our beloved Bonus. He had a recurrence of the illness he had in May and died when it looked as if he might be beginning to recover. He was a gentle and lovely creature and I want to write a little piece entitled "Homage to Bonus" when there's time."

It has had to wait over thirty years. Mary's comments sent by the same post are more revealing:

> "Bonus. This has hit Giles very hard indeed and has taken it out of him – and here as yet we see no hidden blessings. The three days he was ill were a nightmare but I am thankful the nursing was not prolonged and that he went suddenly. Giles has I think given a greater share of his care and love to Bonus than to any other creature or aspect of our life here and I know he looked on it as an investment – training Bonus for a useful and happy future, and we should normally have kept him for two years more."

The financial loss was also severe and since most of our cows were already in calf we decided to rely on AI for the rest of the season and postpone the expense of buying a replacement till the following Spring. When in due course we had to part with Tidy, our next bull, we got well over £500 for him, but he was healthy meat. For Bonus, because he was thought to have died of liver fluke, we got nothing.

As an epilogue to this "homage to Bonus" we quote an account of Tidy's arrival in 1976 from the farm diary which on rare occasions such as this deserted its habit of factual brevity:

Feb 25th
Mr Pascoe rang – after bull sale – no bull for us! Yet! Prices right up – average £700! Trying to buy one for us outside.

Feb 26th
No news from Mr Pascoe so no bull, alas!

Feb 28th
What a day! began with phone ringing at 4 am. Chap at Lealholm – with a bull – how could he get here? M & G struggled into clothes – planning

*rearrangements – moved Jeannie to barn, shut out Krona (horse), blast too late, there she canters into the yard for unexpected company. Halter! Lead her out to Boston and secure gate. Move bewildered Gale from the stable and park her with Auntie Lizzie for rest of night. Find ladder – climb up for bale of straw for stable, get hay but can't find bucket in dark, get it! as enormous van drives into yard. Ten minutes after phone call his lordship's premises are ready. Driver had cup of tea; dazed and sleepy bull happily resting on new straw – back to bed. Why had no-one told us we had bought a bull?*

Act II.
*After breakfast G rang previous owner near Launceston to find out about food, company etc. He was driven/led out each day to drink in yard so M & G put on a halter and led him to trough for drink but then he is startled, starts to trot, bolts, sees windmill, ( red muck-spreader) and attacks it, dumps it, bolts, jumps cattle grid and away over T. M.'s field and through a gap and up bank to the further boundary.*

*Snorting, frightened G stalked him for forty minutes. Talked. Each manoeuvred for uphill advantage! So scene shifted to ¾ way up bank. M cut fence for return journey and finally led out Jeannie as hostage or decoy. Getting interested in J cut down bull's fear of G who was able to start patting and pushing him from rear. V slow progress but with two major diversions procession wound into our yard and back to haven of stable. Rang neighbours twice, once to alert and again to sound all clear. Background noises and sights included hunt and hounds and the fox."*

How grateful we were that Bonus never led us such a dance! He initiated us into the art of bull-keeping. We owed it to him that in the years ahead we were able to cope with Tidy's arrival and many other bull problems as when two trumpeting bulls were advancing on either side of an open gateway and Mary had to run to slam the gate in their faces and hold the fort until help came. But that's another story!

## Chapter Fourteen
## WOODLAND PORK

Pig-keeping lay at the heart of our Bank House venture. From the moment Mary first set eyes on the Nutwood she recognised it as ideal for outdoor pigs such as she had kept for years at her Sussex home. By nature pigs are woodland creatures and here was a wood of eleven acres crying out for pigs to turn its acorns, hazelnuts and much else on or under the ground into delicious food for humans. How else could we put the Nutwood to productive use since the sheep, cattle and horses would gradually destroy its existing trees and bushes besides preventing natural regeneration? It had to be pigs.

At a time when pig-keeping was increasingly monopolised by factory farming the ever more urban public at large met pigs in fiction or not at all. From the nursery days when one's big toe represented the pig who went to market, through to Beatrix Potter's "Pigling Bland" and Pooh Bear's little companion, there was not much hint of the real nature and character of the actual animals until one came to "Animal Farm" in which Orwell got it right about the superior intelligence of pigs if not about much else in their nature. We realised that outdoor pigs would fit in perfectly with everything else we wanted to do at Bank House; with the way we wanted to farm and the people with whom we wanted to share it. Pig feeding is a comparatively simple twice-daily job that anyone can join in with. As almost everyone enjoys stroking a cat that purrs appreciatively, so almost everyone enjoys feeding pigs which so clearly relish their food and their freedom to roam and rootle.

Our pigs would exemplify our preference for mixed farming and extend the variety of experience and responsibility which we would be able to offer our visitors and resident helpers. We liked the economy of making positive use of everything that grew on the farm especially whatever the other farm animals usually avoided, and similarly of all the scraps and so-called waste from our own kitchen. All of which explains why keeping pigs at Bank House was a foregone conclusion, a decision reached without conscious discussion, but the decision, however reached, opened up a host of issues of practical detail which Mary worked on during much of our last year at Abbotsholme where incidentally, she had introduced a breeding sow to the revitalised school farm and trained the pupils and kitchen staff to sort out the scraps into what was suitable for pigs and what was not.

Preparing for Glaisdale she was aware why it might be courting trouble to rely on buying two breeding sows locally once we had moved to Yorkshire, for not all breeds would be equally suitable for the outdoor life

## WOODLAND PORK

we planned. Furthermore, most available sows there would have spent their entire life indoors being bred for quick growth rather than hardihood in a moorland winter, achieving wondrous growths on a mainly meal diet. What we wanted were good scavengers with an instinct for finding food for themselves and these needed a digestive system adapted to a wide range of fresh food and an ability to thrive on less concentrates.

As with our choice of Devon cattle, the quality of docility was also important. Mary knew from experience that the Large Black breed which she had kept in Sussex was suitable in this and other respects. Naturally she didn't want to risk untried breeds that might be less easy to handle, especially in view of the young people we hoped to attract. Although pigs were no longer kept at her old family home she managed to track down the farmer who had bought the last of their herd and arranged that he would sell us a pair of black gilts of the old Madehurst strain in pig for spring 1974.

Their dramatic arrival at the end of January has been described already. We christened them Matilda and Placidia and housed them temporarily in one of three stalls in the stable with geese and the new bull as close neighbours. As all these stalls were needed for more urgent cases during that first lambing and calving season, we set to work whenever there was a moment to spare to build them permanent quarters in the nearest part of the Nutwood. Four fencing posts were driven into the ground at the corners linked at top and bottom by horizontal timbers. Apart from these twelve pieces of good wood the material was very cheap consisting mainly of off-cuts of pine which a timber yard sold as firewood and which we also used for blocking up gaps in hedges. These still had their bark on their rounded sides and were nailed vertically all round as close as possible except for the open doorway. More off-cuts were nailed inside wherever we could see chinks of daylight. For a low, sloping roof we used the sheets of corrugated asbestos that we had removed when re-roofing our kitchen wash-house and these completed a remarkably snug pig house that would not have looked out of place on the set of a Canadian movie about bear trackers. More to the point it was much appreciated by Matilda and Placidia when they took up residence on April 2$^{nd}$ and was still in use when we left the farm sixteen years later having in the meantime served as the prototype for two more cabins which we built to increase choice as to where we could keep our pigs. These additions would not be available for our first farrowings which we hoped would take place early in our second September and for which we needed two separate pig quarters.

Large Blacks on Pig Banks and in the Nutwood

# WOODLAND PORK

In the orchard was a dilapidated wooden hen house, the sort you can stand up in when cleaning it out. We reckoned this would just do for one of the gilts after being improved by a day's crude carpentry. Its lucky tenant would have the run of the orchard and be able to clear up any windfalls before we arrived to provide their official breakfast.

The timing of subsequent service and farrowing had to be precise for a tiny outdoor herd which could not justify the expense of maintaining a boar of its own, so we had to hire one as and when required. If we could manage to keep our two gilts in step with each other throughout the twice-yearly cycle of service, farrowing, weaning and service again, it would reduce both work and expense considerably because, for example, it would make it possible to house and feed them together whenever they were not suckling their litters, which would in fact be the case for all but sixteen weeks in the year. This precision of timetable is based on the unusual feature of pig physiology that the time when you can best guarantee a sow will come in season is five days after her weaning her previous litter, the one somehow provokes the other. Provided we weaned our piglets at eight weeks old we could get our sows in pig almost at once and since their gestation period is a memorable "three months, three weeks, and three days", we had a very good chance of our sows fitting two farrowings neatly into each year with the great advantage of keeping the farrowing to the months that were most desirable, March and September. This had the further advantage that we should avoid having young pigs out of doors in the middle of winter when there would be little tender growth for them as they began to graze, for, yes! pigs will always graze given the opportunity and both the diet and the exercise help to keep them fit and healthy.

That was the plan, devised in advance: we stuck to it all our time at Bank House and never found reason to change it despite all the hiccups and accidents that flesh is heir to. It must be pointed out however that outdoor pigs will plough up all but the hardest soil unless metal rings are put in their noses to deter them. Since our time misguided political correctness has led to banning the practice, no doubt influenced in part by the horrific squealing that accompanies the operation, though an identical noise is provoked simply by lifting a pig off the ground. In each case the squealing ceases the instant the animal regains freedom of movement for, as with many other species, pigs use violent noise as a defence mechanism. The first time Giles had occasion to pick up a piglet – to remove it from a pen for an injection – a sudden, unexpected ear-splitting screech secured its immediate, involuntary release. The political choice is between a momentary piercing of the nose or a lifetime hemmed in by iron and concrete. The arguments for the latter are chiefly financial.

At the time of our coming to Glaisdale small-scale pig-farming was still common. In most cases it didn't pay to keep a boar full time unless the cost could be offset by hiring it out. Many local boars spent much of their lives on tour so that quite often when we arranged to hire a boar we were instructed to fetch him from one of the owner's customers and return him to another when his duty was done. It was all part of a network of contacts which we as newcomers found particularly useful, so it didn't take us long to be on the track of several boars within reasonable distance.

Even at the best of times pigs are not the easiest of farm animals to control or lead about. Our usual practice was to train them to follow some food in a bucket that was rattled or banged with a shovel in front of their noses. It was important for them to learn that there would always be something to eat as a reward for following. Mary used to put their regular troughs in different places so they couldn't cheat by guessing where to go but had to keep up with the bucket. This habit, once acquired, would be useful for the rest of their lives though visiting boars having had no such training were much more difficult to control. Sometimes a boar would be upset by having been removed from its familiar ground and routine and subjected to a bumpy ride in a trailer. On one occasion two of us spent one and a half hours cajoling a boar into our trailer in order to take it home. Bringing new animals onto a farm is always a bit of a gamble. What if this great boar refused to serve our sow? It was with real satisfaction that one day's diary entry recorded: "*Edwin and Placidia entirely amorous – even to exclusion of food!*" On the other hand another passage in a letter was less satisfactory: "*the borrowed boar rubbed himself against some handy stones and brought down four yards of the orchard wall.*" He was a great beast!

Of course the procedure could be reversed so that the lady was taken to the gentleman, which happened if we didn't want to wait for the boar to finish his work elsewhere. Usually the boars were fashionable Large Whites which when crossed with our black sows produced white piglets with random, often comical, dark patches normally referred to as "blue". On rare occasions we needed to replace a black sow and then we had to seek much further abroad for a black husband. Once, the nearest available were at Coventry, Norwich and Whitehaven. We plumped for the last so Giles could spend a couple of nights with his sister outside Kendal. On that occasion securing a black litter involved two long days of slow driving with a trailer. Another time Matilda had to be taken for a sixty mile ride to beyond

## WOODLAND PORK

Selby. That trip illustrated another hazard for when Matilda was left with the boar for a while neither showed the slightest interest in the other and the owner told Mary to take Matilda home for the night and return next day. He explained that his only accommodation for her was on the other side of the main road which happened to be a very busy, fast stretch of the A19. He hadn't bargained for a tame sow. When he saw Matilda obediently following Mary's rattled bucket through a labyrinth of gates and passages in order to get back to our trailer he relented. "If it's that easy she can stay and I'll phone you when she's served and wants fetching." Matilda stood close to Mary and her bucket of food, waiting patiently for a gap in the traffic and then trotted across in perfect safety.

Most people attracted by the principle of keeping pigs out of doors have little idea what it involves in practice. To begin with it requires sound fencing or other containment to all the areas the pigs will occupy. The majority of farm hedges would not deter curious or hungry pigs for long, and most pigs seem to be both curious and hungry most of the time. The mature are immensely strong, shaped like torpedoes and apparently impervious to thorns; the young are small enough to slip through holes left by rabbits, and all ages are ready to capitalise on electric fences broken by cattle or with run-down batteries. Stone walls in good repair were best, otherwise we had to provide strong pig-wire fencing with the bottom wire let into the ground or pegged down to it. This explains why we concentrated on fencing right round the Nutwood during our first two winters.

Fortunately for us imported boars rarely took advantage of their greater height, weight and strength to break bounds, perhaps because they had other things on their minds, and were largely content to follow the sows. It was also fortunate that when piglets were at their smallest and could penetrate our defences in many places they were anxious to stick close to their mums and when by mistake they got into the wrong paddock they usually found their way back. By the time they were so to speak old enough to go to school and took to roaming in gangs, they were already getting too fat for their earlier, narrow escape routes. The really testing time for our defences came with farrowing and weaning when the sows' strong maternal instincts could not be contained easily. Generally farm animals very much dislike living alone and most of the year our pair of breeding sows lived together happily even though usually one was boss. However, when it came to farrowing time they wanted to be on their own. As with the cows and sheep, one of the surest signs that the hour of giving birth was imminent was that the mother to be went apart. Pigs can demonstrate an almost human perversity. One day a sow might break out because we had separated her from her buddy too soon. The next she might break out in order to be alone. Only very close observation could get it right but we couldn't sit around all day to be sure of that.

There are many signs of imminent birth common to mammals in general such as what is colloquially known as "bagging up" as the burgeoning milk supply makes itself evident. Given the freedom to do so pigs can tell us by another, unexpected way which is virtually infallible. Five hours before farrowing our sows would begin to make nests. They appeared with great mouthfuls of material, grasses, rushes, bracken, heather, sticks and even the odd length of honeysuckle and these they assembled – you would hardly call it building. It seemed to be little more than a rudimentary, symbolical activity. The purpose of this vestigial behaviour was made clear to us by Hilda, Placidia's successor. We planned for her to farrow in the house on Pig Banks and had provided a generous bed of dry straw for the occasion. At her supper time she was missing. Eventually she was discovered in Black Carr having let the house cows follow through the gap she had created in the fence. She was busily engaged in making a bed in the middle of the biggest patch of rushes. Besides driving her back to the Pig Banks house and securing the fence, we tried to humour her by carrying armfuls of her rushes up the steep bank to add to the straw in her house. Fond hope! On a late visit to check that all was well we found her back in Black Carr. By dusk she had made a huge mound of rushes, six feet in diameter and three feet high, and disappeared beneath it. We admitted defeat and left her for the night.

Next morning, to quote from the diary:

> "a gang of photographers attended on Hilda and her 'house' of rushes. Her back end and an unknown number of piglets were visible. Later her head appeared t'other side and piglets used her tunnel: apparently nine of them."

When it drizzled the following morning mother and family remained perfectly dry as the rounded mound left the top-most rushes sloping outward and downward in a most effective thatch. For how many generations of indoor pigs had this instinctive, unpractised architectural skill lain dormant? How much can animals teach us if only we let them! On the other hand those same pigs can blunder. More than once when a hot dry August seemed to have no end our sows would judge it safe to farrow in most unsuitable places. Twice Matilda selected a dry ditch at the very bottom of the Nutwood only for a thunder storm to turn the hollow into a stream. By the time we discovered her she was past being moved. Giles remembers fetching posts and sheets of corrugated iron and standing in a torrent of summer rain to nail together a crude shelter over Matilda as she produced one piglet after another, and then slopping about with bales of straw to improve on her collection of oddments serving as a bed. Somehow those piglets proved none the worse for their premature baptism.

The birth of a lamb or a calf is at times complicated and dangerous if the creature to be born is unduly large, has a leg doubled back, or is lying with its body back to front instead of preparing to issue head first in the ordinary way. Every shepherd or cowman meets with such abnormalities often enough to become skilled in coping with them. Not so the pig keeper. It is comparatively rare for a traditionally kept sow to have much difficulty in the actual delivery of her litter. Relative to their large mothers piglets are tiny and can pop out easily without the familiar signs of prolonged straining endured by ewes and cows. The combined weight of an average litter – about ten in our case – may well be as great in proportion to the sow's weight as that of a promising newborn calf to its mother or twin or triplet lambs to their ewe, but since that weight is made up by some ten separate piglets the delivery of each one in succession does not usually present any undue problem.

The one exception in our experience followed an accident when Placidia got badly squeezed well on in her pregnancy. The subsequent farrowing was related in a letter:

"First then Placidia. Her birthday – so to speak – began with dismal promise and suspense; two dead and two alive and no more for hours. Ought we to "assist" or get a vet, or what? Mary rang Bill who had kept pigs for years, but he had never had to help with a pig as one does with sheep and cows. So a watching game in relays followed and much pessimistic guessing. However after a five hour gap she resumed production unaided and out slithered another eight piglets one of which she rolled on in the night, so we are, or rather she is, left with a healthy family of nine, which is better than we'd dared hope for."

The danger of a sow lying on her own newborn piglets seems to be nature's way of adjusting the balance between her over-generous fecundity with the lesser needs of maintaining the race. The wonder is not that so many expire but that so many survive. It is fascinating to watch the care with which so massive and heavy a creature lies down amidst her seething progeny without disaster. As her bulk gently subsides, often accompanied by an expressive "oomph!" of expelled air, a piglet is bound to dart out of safety like a city kid dodging between moving traffic for fun. It was always thus. Shakespeare brings on Falstaff, preceded by a small pageboy, saying "I do walk before thee like a sow that hath overwhelmed all her litter but one!" Fortunately we never sank to that level!

Having survived the hazards of birth and infancy the piglets would grow apace on their mother's milk, provided her milk bar was working on all cylinders with a teat per piglet. Until we weaned them at eight weeks the piglets ran with their mums and attended their trough feeding of meal plus scraps. Gradually the young noticed what the adults were doing and began to poke in their noses in exploratory fashion. One day one piglet would start to eat for itself, then another, and another. Once they were all eating steadily we transferred their meal into a creep which we had built that September as we entered on our second year. This involved constructing a concrete slab – under which, incidentally, we were able to bury a lot of the broken glass and tins our predecessors had heaped up in the Nutwood – and fencing out the sows so the piglets would eat their rations at their own pace and not watch the adults gobble them up instead. In subsequent years, when we had built other houses and feeding slabs, it was instructive to compare the litter that was kept in a field with little more than grass to add to their meal diet, with their Nutwood cousins who were much later beginning to eat meal seriously, preferring to find fresh autumn food for themselves, especially the hazel nuts and acorns which abounded. In spite of their delaying to eat the proffered meal long after their cousins were tucking into it on their grassland, the Nutwood foragers grew the faster.

One of these particularly well nourished Nutwood piglets was Hilda whom we have already met as the architect and occupant of an igloo of rushes. When Placidia died prematurely Hilda was reprieved from being despatched to market as a weaner and kept on as a stop-gap replacement to breed from. Naturally Matilda didn't readily accept Hilda as her companion of house and trough, so Hilda spent a lot of time on her own, compensating for her lack of pig company by attaching herself to people. On occasion, visitors being taken on a tour of the farm, were surprised to find a tame pig in attendance. Our Christmas letter that year reminds us vividly:

> "Hilda, breeding sow elect, choosing the vacant Black Carr Buildings as her summer residence from which to sally forth at will and grace all our field work with her presence, arriving triumphant on rocking-horse legs with syncopating ears."

> She would take full advantage of a series of gates left open to facilitate muck-spreading, and explore widely. Her joy on coming across one of us fencing somewhere soon palled so she took to teasing. The fencer, kneeling down to fix a bottom wire, with a couple of staples between his lips, would

## WOODLAND PORK

put out a free hand for a hammer only to find that Hilda had moved it out of reach. This happened repeatedly so we became convinced it was not chance but a deliberate game, an explanation that seemed all the more credible in the light of piglet behaviour we had been delighted to observe and described for the benefit of Giles' parents.

"I've discovered that they originated musical chairs, though of course in their version they provide the music themselves. The rules of the game are based on troughs with at least one more compartment per side than the number of players. At the word go the umpire dollops in food and all piglets set to. As soon as one is convinced that somewhere else there is more food than in his own section he races round the block looking for an empty place, thus vacating his original position which immediately tempts someone else to move in on a similar assumption. The aim is to make sure that as the food in any section is genuinely and entirely consumed it is not the one in your possession at the time. Players found without a trough section containing food and unable to find alternative accommodation have to drop out until there are only two contenders left for the last remaining visible food. The stronger declares himself the winner while all losers go off to inspect empty buckets in hope, to chew the umpire's gumboots in case they might qualify as food or to follow their mamas to find nuts or grass, pretending that they weren't really playing anyway. The speed and urgency with which they circle in trying to find a new opening is very comical.

Incidentally, Hunt-the-Thimble and Sardines are clearly among the many other games invented by pigs and copied by humans, as was of course Piggy-in-the-Middle. I might add that men must have learned wrestling from cats, King-of-the-Castle from lambs, Skipping and Hide-and-Seek from calves, and Jumping from horses. Winsome plays Grandmother's Footsteps with our garden cabbages."

When we began keeping pigs at Bank House we expected to sell off all our weaners at eight to ten weeks old rather than fatten any of them ourselves for pork or bacon almost entirely on bought-in pig meal. November 1974 found us with Matilda's and Placidia's first litters weaned and ready for sale so we ventured forth for the first time to try our luck at selling and took all twelve of them to Ruswarp mart. To our surprise we found we had stumbled into the Annual Sale and Show. Alas as far as prices were concerned the general verdict was that it was a "poor sale" so that although we were heartened by winning first prize we came home with a cheque in our pockets for only £107. The following autumn though our weaners were not noticeably of better quality, prices were up again and we got £24 for each of them. Some you win, some you lose!

In our third year, with prices back down again we were beginning to find time to look about with more experienced eyes and to be more flexible and opportunist in our pig selling. The unchanging basic fact remained that grassland hill farmers will never become rich selling weaners fattened on bought-in food; yet there were various chances that came our way of selling one or two pigs from home and we made more from each such transaction whenever it occurred. In the first place someone who was impressed by the evident health and vitality of our weaners and wanted to buy just a few of the gilts among them to breed from, would be ready to pay above the weaner flat rate. So was the couple running a caravan site who bought a pig from us each summer to eat up all the scraps and waste food. Not being farmers they preferred to deal with us rather than bid in the open ring of the professional market. The pigs we sold them became a sort of mascot of the site and pets of all the caravanners. For some reason they named all their pigs "Sam" regardless of their sex. We also began to keep on one or two weaners from each litter with an eye on the Christmas market: those born in early September would be ready as pork whereas those born in March would have had time to mature into bacon. When a family connection ordered half a pork pig in advance we were able to sell the other half to our village butcher and this was distinctly more profitable than its weaner siblings had been. Then we began to cure our own bacon, salting whole sides in large, old-fashioned oblong porcelain sinks kept in the cool damp dairy. We finished these off in the warmth of the kitchen. The thick hand-made iron hooks we had long noticed in the kitchen beams were once more used for their original purpose and for several months a year all but the shortest had to duck beneath these great flitches when moving around the kitchen. As with the pork so with the bacon. The first two we cured were shared between ourselves and the butcher. Whatever we consumed on the premises had the twin advantages of being cheaper than shopping for meat elsewhere and a source of satisfaction. Whereas much bacon on the market produced water in the frying pan, our home-grown, home-cured bacon produced lots of delicious fat in which to cook the eggs as well. Though we said it who shouldn't, our bacon was as good as the best in the land, and many customers agreed.

## Chapter Fifteen
## HORSE POWER

For centuries heavy work on the land involved the use of oxen, horses and donkeys until the agricultural twin of the industrial revolution substituted machinery. The widespread historic change from draft animals to tractors was largely complete in the North York Moors by the time we came to Glaisdale. Though isolated pensioned-off horses were to be seen grazing quite often, few did any regular work. Oddly enough the capacity of car and tractor engines was measured in terms of "horse power" though the actual horses had gone, leaving a double legacy which we were to find extremely helpful. In the first place there was a remarkable range of redundant horse-drawn machinery at farm sales together with an even more varied assortment of harness that would be essential for its revived use, all of it having been gathering dust and cobwebs in the deeper recesses of farm outbuildings. Equally important to us was the great reservoir of remembered experience, the detailed knowledge of how everything worked and how to get the horses to cooperate willingly and effectively, because all the elder generation of dales farmers among our neighbours had been brought up using "horse power" literally. When Bank House had last been competently managed back in the 1950's there had been work for eight horses. Alas, our predecessor possessed neither horse nor tractor. We planned to own and use both.

There were several reasons for our reverting in part to an outmoded system. We were naturally influenced by Mary's experience at her family home in Sussex where she had inherited a pair of chestnut cobs. With these each year between 1945 and 1952 she had cultivated three acres of root crops, made thirty acres of hay, mucked out a fold yard and occasionally re-seeded some pasture under- sown with oats. She had loved the work and looked forward to sharing some of her thrill and pleasure with young people at Glaisdale. Less subjectively was the sober recognition that many possible helpers would have little inclination or aptitude for matters mechanical and we wanted to be able to offer a variety of activities to suit their differing interests.

The lie of the land also supported the decision to use a horse since a considerable part of the farm is inaccessible to tractors. Furthermore, our interest in achieving a degree of self-sufficiency made it more attractive to us to grow the food we should need for a horse than to rely absolutely on diesel fuel from abroad, a consideration much on our minds during the world-wide oil crisis which led to the issue to us of petrol ration coupons just when we were starting up. Fortunately they were never used as the crisis was resolved, though a possibly related event followed a few years later when an international company began prospecting and sent massive echo-sounding vehicles thumping up our drive and way up the bank above. What if they were to discover oil instead of potash?

So there was another matter demanding our attention during our first eighteen months at Bank House to add to our search for heifers, sheep, pigs and a bull, not forgetting a sheep dog, all competing with the priority task of building barns and fold yards. Where could we find a suitable horse?

Besides consulting neighbours and everyone we met likely to be in the know, Mary made time to attend marts and shows, beginning in our second week with the great annual show at Egton when fifteen thousand gather on the hilltop opposite Glaisdale. When she phoned the secretary of the Dales Pony Society she was told that there was a grey gelding available not far away at Moorsholm. We didn't follow this up because Mary was set on buying a mare from which to breed a working companion and ultimately to train her own pair for mowing and other field work. The search was prolonged over the course of a year taking her further afield; to Harrogate, to Darlington and to the Dales Pony Show at Bowes. She discovered that in North Yorkshire and County Durham there were still a few of the black Dales Ponies traditionally used for shepherding and light farm duties. Eventually she tracked down a mare of a similar type, a Fell Pony said to be "broken to harness" and relatively cheap because she was too big for showing. Mary went to see her and reported that she was "highly desirable and equally expensive". After more trips to view her and lengthy negotiations we finally agreed to buy her. It was February 5$^{th}$ 1975 before our diary was able to celebrate:

*"Landmark day. Mr Harker rang – was the lady about?" So he brought Krona in the morning and a new face looked out of the stable when the rest of us returned for lunch.*

From the word go a new element of excitement and responsibility entered our lives, a mixture Giles conveyed to his parents as soon as he had time to write:

*"Krona, big black "pony" –i.e. horse, is here! Mary got the creature here and then went off to Leeds for the night leaving us to cope – i.e. mend fences and retrieve! She can open most doors, reach over fences in devastating fashion, and likes to get near other animals who seem to think she is possessed of the devil."*

Mary added a note that was more complimentary:

*"Actually she is delightfully sociable – comes when called, accompanies anyone crossing her field and is superbly beautiful."*

That first afternoon we put Krona to graze in the nearest field along with Bonus, Jeannie and Duchess but her well-intentioned social sallies sent Bonus over two fences. On Day Two this little posse of cattle were still panicky and cowered in the yard all morning, so we put them to graze in the next field. The diary for Day Three began:

*"Krona not in sight first thing! Discovered eating Mr M's grass – easily brought back but found to have exited by walking through the boundary fence."*

Day Four's diary was better news:

*"Bonus jumped into Barley field to be with Krona. Great surprise and progress. Jeannie followed. Peace established. Panic over."*

That was only the beginning. There was a very long way to go before we could expect this beautiful, wilful creature to feel so at home with us, the farm and its other denizens that we could get her to perform the hundred and one tasks we were lining up for her. Always at the back of our minds loomed the uncertainty whether we should be in time for Krona to contribute substantially – or at all – when we should need her most urgently, as we faced hay-making with our single tractor and minions with hand rakes and pitch forks. For a month or so success seemed unlikely and for several reasons. In the first place we still lacked necessary implements and harness, and since acquiring them depended more on chance at farm sales than on a liberal cheque book there was little certainty. The calving and lambing seasons were well underway so it was extra hard to find enough time for training Krona and getting her used to the unfamiliar demands we made on her. She needed much more handling and human company than she got and as it couldn't always be Mary she had to tolerate the fumblings of some of us who were only too obviously novices, but we too could learn, and we did.

The first step we could take to establish a working relationship with Krona was simple riding. For this purpose on Day Four we bought basic harness locally and cheaply: a bridle for £6.07; a girth for £1.48 and stirrup leathers for £2.33. We already had a saddle which had come from Sussex with the pigs the year before so the next day Mary was able to take her first short ride, which Krona celebrated by letting herself into the front garden three times, apparently hoping to enter the house. Others took to riding recreationally, usually when leisure and good weather coincided; first round the farm, then venturing in various directions especially into Arncliff Woods and onto the edge of the moor. Mary rode to a hunt meet just along the dale: the saddle slipped when Krona got excited to be with so many other horses and Mary came off. Several of us had similar mishaps without as much excuse but fortunately none of us was hurt. The diary reminds us that one ride nearly ended in tragedy, on a dismal wet November afternoon:

> "S, O & T took Krona on moor. Drama! Went into bog right up to her middle – stuck! T. rushed down for help. M, G & T took up ropes, planks etc in dusk to find O & S had just got her free."

They had panicked her into pulling herself out. In total contrast was the ride undertaken by Sally just after leaving us, a twelve day trek over the North York Moors which culminated in riding into the sea at Runswick Bay. What better way could there have been of making a horse fit for hay time?

Gradually rides became more purposeful, carrying out necessary jobs such as the daily inspection of sheep and cattle, then rounding them up and moving them. Or, occasionally, taking an important letter to catch a particular post or picking up medicines left by the vets for us to collect from other farms. Before a month was up Krona was becoming distinctly useful. In particular we were finding what a valuable farm asset she could be as a pack-horse, sometimes ridden sometimes led. All sorts of burden were more easily carried by her than by us foot-sloggers. There were sacks of chicken manure to be spread by hand, ailing lambs to be brought in for nursing and, of course, hay for sheep wintering in remoter parts of the farm. A pair of our light, loose hay-bales would be linked and slung across the saddle to hang down either side in balance. A rider could also hold a third bale on his or her lap, and we could even add small bags of sheep nuts if there were help at hand to deal with gates.

The first implement of any sort we were to try out on her was a crude bush-harrow for her to pull over fields of short grass. It was constructed of thorn branches lashed side by side at their stout ends to a cross bar to work like a five-foot-wide besom lying flat. It was likely to wear out easily and would require frequent repair but there was no shortage of band or branches. It would be useful training and the hedges would get trimmed. Sally had joined us specifically for the opportunity of horse-work and she it was who made our first bush-harrow. She and Mary set about training Krona to use it.

## HORSE POWER

It must be frightening for a horse the first time it is yoked to something that follows her whenever she moves, especially since every attempt to get away from it by accelerating or changing direction is immediately copied by her pursuer. Early lessons with the bush-harrow therefore needed two people in attendance to control and calm her. Even though by now Krona was getting used to being handled, this uncanny devil behind her proved too much. She took off straight for the closed gate she had just been brought through and leapt at it. She cleared the gate admirably but the harrow did not. End of bush-harrow mark I! Thereafter perseverance paid off and within a week the diary recorded:

*"Successful session training Krona to pull brushwood harrow – only one serious attempt to bolt, just thwarted. Still nervous but a great puller."*

A week later we began to manage with only one attendant once she had started off. This progress with the bush-harrow was paving the way for much more important work.

In mid-April we had had a stroke of remarkable good luck. We had gone to a farm sale at Liverton cherishing hopes of buying something advertised, only to be disappointed when the bidding for it soared well beyond our means. Looking around the field we spotted in the distance what seemed to be a heap of scrap metal. Closer investigation proved it to have been a horse-drawn side delivery rake, rusty, but more or less complete if reassembled. A scrap iron merchant, the only bidder besides ourselves, soon dropped out, so for a mere £6 we bought one of the most valuable assets to our hay-making for years to come. Describing it to his parents Giles wrote:

*"It looks as though it were invented by a weaver, having four horizontal bars of hanging tines that move sideways in insect fashion, an improbable, un-machinelike motion, moving the hay from right to left but with a freedom to make your new rows in whatever curves or straight lines you like."*

Normally it turned the hay over onto dry ground between the rows. If the hay was dry enough the operator, sitting high above the combs of tines, could choose to amalgamate the two rows into a bigger one ready for the baler. At that time we couldn't afford a second tractor but we now owned a combination of machines for tractor or horse, to mow, turn, row up and bale our hay, and if one process of the assembly line could be performed by a horse it would  free the tractor to work simultaneously, both forms of "horse-power" in tandem.

Krona and the side delivery rake in action

## HORSE POWER

Without Krona's contribution our hay- making would be repeatedly delayed when the only tractor was changing over from one implement to another; only one process could take place at a time and only one person was being employed. In typically unreliable English midsummer weather, these limitations often made the difference between getting hay made and not. Provided we could persuade her to pull this heavy ironwork for as long as it took to turn a whole field we should have solved the bottle- neck of our haymaking procedure.

With haytime rapidly approaching we had much to do if the side delivery rake was to play its part that year. We had to buy plough-lines, extra long rope reins. Shafts were fitted on the rake which was man-handled out into the space of an open field where it could be tried out more safely. Krona was ridden more frequently and work with the bush-harrow continued from time to time. Her education was shortly to be further extended by pulling iron harrows in a field of newly sewn rape seeds. We experimented with her harrowing the brittle young bracken shoots in the top fields, the "harrow" in this case being a gate or an old bedstead. Though they were not sharp enough to be effective it was useful experience. At last on July 2$^{nd}$, with fingers crossed, Sally drove the side delivery rake into the lanes of drying hay and kept going until a third of the field was done. Next day the rest of the field was finished and Krona's haymaking career was launched.

Turning and rowing up were not Krona's only contribution to our haytime. We had had two other bargains at farm sales, both out-of-date horse-drawn implements. One was a different sort of hay-turner with two sets of tines whirring round in circles; the other a rake with a row of large curved tines that were raised and lowered together for clean raking whatever the baler left behind. Since we needed every scrap of hay in those days we were in the habit of clean raking every hay field, so it was a real advantage to have it done by a horse while the tractor was mowing or baling elsewhere. We could, and sometimes did, use the tractor to do it with the Vicon Acrobat, depending on who was free at the time. Mary was delighted when Giles said the horse rake did a much better job than the Acrobat in everything except speed. The rake missed much less and left what it collected light and fluffy whereas the Acrobat created a rope. An added advantage was that it could be operated by someone with lesser horsemanship than was essential with the side delivery rake.

Krona Harrowing and Raking

## HORSE POWER

The relative merits of horse and tractor were often debated at mealtimes. Each had its devotees, but the pleasure of working in the manner of one's preference carried as much weight as strict logic – which was as it should be amongst volunteers. Mary put her view in a letter:

*"I had forgotten the great pleasure one has always derived from driving a hay turner, perched high up on a springy seat, able to see the rakes whirring in front – no neck craning as with the tractor: no noise!"*

Krona too found hay work enjoyable, was pleased to be fetched to work and thrived on it. *"She is now black and shining and fat and glossy."* An indelible memory of Bank House for many visitors as well as resident helpers and local friends was the spectacle of a horse with her long tale and mane flying and Mary mounted on the rake behind her, using the momentum of acceleration to help her up a slope, and with hay pouring out sideways like a fountain to exaggerate the impression of speed. Even standing still in front of you Krona was a formidable sight. Her great muscular shoulders and broad chest gave an awesome sense of power that made you think of cavalry charges or of Boadicea and her war-chariot.

Haytime provided her with the longest and most intensive period of work of her year but once the side delivery rake was stored away for the winter there was still a great variety of occasional jobs waiting for her which were quite unsuitable if not impossible for a tractor. Many of these jobs were accomplished by another reversion to a method of transport once commonplace in the North York Moors but almost entirely abandoned by the time we arrived: a horse-drawn sledge. For this we were indebted to Bert, a son of a local farmer. He had the imagination to realise that a strong horse sledge would fit in perfectly with the rough terrain of Bank House and with our way of farming with young folk. Fortunately for us he was at a loose end so he proceeded to design and then make it for us. It was simplicity itself consisting of two beams about seven foot long, three inches thick and fifteen inches wide, set on edge to act as runners. A superstructure of stout members formed a sort of intermittent platform fifteen inches off the ground. The runners were rounded at both ends so the sledge could be pulled in either direction, thus saving a lot of trouble when heavily laden, since it could be reversed easily instead of being turned round. Bert managed to procure and fit very strong steel strips to protect the underneath of the runners from undue wear on rocks and stony tracks or yards, and to make the sledge slide more smoothly.

The sledge was ready for use on February 19th 1976. It took Krona very little time to learn to pull it, her only serious difficulty was going downhill if the sledge threatened to run into her from behind. At first this was cured by attaching short ropes for attendants to hang onto from behind as human brakes. Later Bert provided loops of thick iron chain which could be slipped over and under the front of the runners. The sledge could be hauled where it would have been foolhardy or downright suicide to take a tractor; up steep banks, across sideways slopes, over bogs and amongst trees. Being low it was much easier to load than a higher trailer, especially with very heavy objects such as stone gateposts. With its crossbars it was very simple to lash down great piles of firewood, bags of sheep and cattle nuts, sheep troughs or bales of hay and straw. It also provided exciting joyrides for passengers, most often on return journeys. Its main limitation was that usually Krona needed someone on foot to lead her by her bridle going through gateways. However the sledge was always ready at short notice and available for service in all weathers.

Though we have memories of using the sledge on many occasions and for many different purposes, two are particularly vivid. One is of fetching five-foot high loads of bracken which we had cut and dried high up on steep banks. We got more and more ambitious trying to see how much could be stacked and roped without the load slewing off sideways as the sledge bumped its way either to a tractor and trailer waiting below on level ground, or all the way into Black Carr Buildings where it was wanted for bedding. Sometimes the sledge could only clear the bracken from high ground by zigzagging like a slalom course both up and down. Mary was fascinated that after being led very firmly the first time Krona anticipated the sharp changes of direction and sometimes reminded Mary if she forgot.

The most dramatic use of the sledge was in deep snow. Krona seemed to enjoy pulling the sledge at any time but when snow made it easier she was in her element. One year we had to cope with deeper snow than ever before or after. It was February 1978 and with still only three inches of snow we had to take the car to Whitby in the dark to fetch a lad who had got the wrong train. On the way back the car couldn't make it up the drive and had to be abandoned. Next morning there were ten inches of snow everywhere. Not wanting to risk stranding the tractor, we took hay and nuts to all the sheep on the sledge. There were different parties of them in outlying fields where we kept them in midwinter, preserving nearer grass for lambing time. Though none of them seemed distressed and all of them tucked into the hay and nuts healthily they were finding movement difficult and their digging for grass no longer effective, so two visits a day with the sledge became the regular pattern until additional light falls of snow made it a foot deep when we could no longer wade through it fast enough to lead Krona and Krona

# HORSE POWER

couldn't forge her way slowly enough to stay with us. Instead Mary sat on the sledge controlling Krona with long plough-lines which allowed her to dash at the snow with a marvellous high-stepping action and much increased speed. Sturdy outriders carrying shovels were necessary to dig out the gate-ways but they too got a free ride some of the time.

Worse was to come. On 13$^{th}$ of the month – not a Friday – we woke to find about two feet of snow in the garden and the garth and twenty inches over the lower fields where the sheep were. A reconnaissance revealed that the ewes could hardly move, and the gimmer lambs not at all. Somehow we had to bring them in. Those of us who had been accompanying the sledge on foot had found it much easier to walk in the ruts in the snow left by the sledge. Couldn't the sheep do likewise? Yes, they could! Which explains how we still have some remarkable photos of a sort of Pied Piper scene with a very long line of sheep, single file, nose to tail, following the distant, hooded figure of someone – too small to say who – carrying some hay as a lure, winding its way through the lower fields and up the S bends of the bank into the farmyard. That night, thanks to Krona and her sledge the whole flock was safely under cover, every building being pressed into service. Krona was put into a vacant calving pen so that sixteen refugee sheep could occupy the vacated new stable which very fortunately we had just finished building five days before. This stable was created as part of a major improvement to Black Carr Buildings. It had doors at either end; the outer ones were nearly always pegged back open so Krona could come and go as she pleased between Black Carr field and the building. The stable's inner doors opened directly into the main covered stockyard. In years to come she would spend many an hour hanging her head through the open upper door supervising the herd very responsibly. Krona was astonishingly hardy and preferred the stars over her head to any building even in deep snow so her stable was often available as an additional calving pen, but never again to house sheep.

A stout farm cart had been sent up to Bank House from Sussex with the pigs our first winter. Mary had used it there in the past for carting turnips and muck but here we never used it much, chiefly because of the steep banks and rough ground. For its first year we had had no horse, then when Krona came we collected bracken in it until a shaft broke and it took us a long time to persuade a retired wheelwright to renew it.

Snow Rescue

# HORSE POWER

After this the sledge proved more appropriate partly because it didn't need such a skilled driver. Mary entertained nieces with it driving to Church one Sunday when the car wouldn't start. Crossing the dale involved a stiff pull but Krona loved the gentle undulations of the road to the village and felt cheated if not allowed to trot. They tied up by the old ring in the wall of the vicarage stable that had probably not been used for a long while. Mary remembers that Krona remained content for fifty minutes:

> "But during the last hymn I heard her whinnying and went out to tell her we were coming. She was so impatient to be off she wouldn't stand still to let us climb in until the Church Warden gallantly convinced her of his authority."

Visits to or from a blacksmith were occasions of great interest. Sally was to take Krona for the first time and it looked as if the shortest route lay over the moor, avoiding quite a lot of hard road. But how should they find the way? Mary had been given the answer. "Start up that track. At the top you'll see a stone on the moor ahead of you. By the time you reach that you'll see another, and so on." And Sally found it so. Krona was involving us in an ancient local culture established long before the coming of tractors. Our life and work in Glaisdale was all the richer because we belonged to both traditions.

Part of our long-term plans for Bank House was to breed a companion from Krona and train it up to form a working pair with her. Soon after her arrival Mary set about finding a suitable stallion which was a difficult task if the resulting offspring was to be well-matched with Krona, and in any case stallions were few and far between. Eventually we heard of one in Richmond and arranged with the owner to leave her there for a month. The stay was extended to seven weeks which was not perhaps, in hindsight, a good augury. As the following summer developed there was little sign of a foal. One day – it was now July 1976 – Krona was "carrying on in no uncertain manner" as the diary put it, and Mary was convinced that far from being in foal Krona needed serving that day. In a fit of frenzied organisation Mary persuaded the reluctant Richmond owner to accept Krona back for just one night, borrowed a horsebox, got our garage to fit a different towing knob on our car, manoeuvred Krona on board and drove off. They returned next day apparently having been successful.

Another long sequence of months passed as we waited impatiently. Local interest spread. We were visited by neighbours including a retired farmer who lived opposite us across the dale. He watched everything we did and often offered sound advice so we had nicknamed him *"The Advisor"*.

# HORSE POWER

The general verdict was that she was not in foal. Another season gone! Mary attended the Dales Pony Society Show at Bowes and made new contacts which led to her taking Krona off in another borrowed horsebox to a stallion near Barnard's Castle: once again apparently successful. Another eleven months' wait ensued, the period of gestation, with more visits from "*The Advisor*" and other interested parties, expressing diverse opinions. At length however signs were more hopeful. While remaining so slim it hardly seemed possible that that sleek belly could contain a foal, Krona began to bag up ever so slightly. Naturally we kept close watch as we went about the normal busy routine of the farm. We were having tea late one afternoon when the phone rang. It was the " *Advisor's*" granddaughter. "*Granddad says to tell you the mare has foaled and they're both well!*" Forty minutes earlier Giles had stood beside Krona and asked "*what about it ?*" and getting no answer went for tea. Across the dale through a pair of binoculars "*The Advisor*" watched the event we had waited for so long. It was May 19th 1980.

## Chapter Sixteen
## HAY HARVEST

Hay-making, or hay-timing as it is commonly known in these parts, was the very heart of our year's pattern of work at Bank House. Other major projects and individual improvements might be brought forward or delayed if we wished but hay is a hard task-master, it has to come first when it is ready and that is not only when the proverbial sun shines.

The growing season in these northern counties is much later and much shorter than in Sussex where Mary had gained her farming experience, a fact we had to come to terms with very soon. So we asked our neighbours when we might expect to have to bring our cattle in for the winter on an average year and also when in the same mythical average year we might expect to be able to turn them out to grass again in the Spring. We learned that the traditional dates were somewhere near November 12$^{th}$ and May 12$^{th}$ respectively, giving us a distinctly longer period of winter regime than we were used to, and consequently requiring a correspondingly greater store of hay. Of course money could buy in hay or alternative food for our cattle and sheep in winter if we had it, which we didn't, but in any case we didn't wish to buy it except in an emergency because hay-making seemed the most suitable and attractive work for us to offer our summer visitors and longer term resident helpers. Our need for the hay we could make on our own land and with our own hands was urgent and genuine and gave validity to our efforts. It gave satisfaction as we worked in the fields in summer and again as we fed it to our stock in winter. It also created a communion of purpose, a sociable bond of team work in which labour was often enjoyable and rewarding.

The hay season cast its shadow long before we reached the symbolic image of bringing in wagons loaded with bales of sweet-smelling hay. In fact planning for the following year began almost as soon as the previous season was over, while recent difficulties and observations were still fresh in mind. It was also inspired by what we had been noticing of the efficient practice of our neighbours; not that we were either desirous or capable of copying them closely. In two particular respects we chose not to follow fashion or to try to keep up with the Joneses. One was our considerable reliance on hand work and hand tools: the other our use of Krona. Both contributed substantially to our purpose of providing unskilled young people with appropriate, even therapeutic work.

There were three distinct aspects of preparation for the hay season. First there was drawing up what might be called a plan of campaign, deciding in how many and in which fields to make hay that year. In our first year poor fertility and our lack of past experience led us to attempt haymaking in every field we could spare from grazing, so we found ourselves in some awkward situations on account of boggy ground and steep slopes. We also had to plan the order of tackling these fields well in advance so that they could be "laid up" (cease to be grazed) in good time. We soon learned that in these parts it was highly desirable not to graze a hay field after May 1$^{st}$ in order to allow time for the grass to grow. Fields yielding the best crops were usually laid up much earlier while those grazed after bore the poorest.

The second preparatory work consisted of steps taken to improve the fertility of each hayfield. Instead of applying the heavy doses of artificial fertiliser for immediate effect which most farmers relied on we concentrated on long term benefit of having forty tons of basic slag spread during 1974 and a lot of lime in 1975. But the greatest contribution to improving fertility was made by spreading the muck accumulated in the foldyards of Black Carr Buildings during the previous winter. The amount available that Spring was very small because the herd had been able to wander outside freely, and because we had not bought in straw as litter. Thereafter we did buy in large quantities of straw every year and greatly increased our muck. That first year we had so little muck to put out that we couldn't spread it over whole fields. Instead we had made maps during haytime recording the least fertile parts of each field so that what little muck there was we put where most needed. In one respect it was just as well there wasn't a great deal more muck that year; since we had no fore-loader at the time we had to shift it all by hand using gripes. The deficiency was remedied that Autumn at a farm sale when we bought an ancient cab-less Fergie tractor complete with a rather frail fore-loader – a hugely popular landmark in our progress. Clearing the muck from the foldyards each Spring was an essential preparation for haytime since until it was done trailers laden with new hay couldn't be drawn alongside the barns for unloading.

The third main area of preparation for haytime concerned the tools and machinery; exploiting the Autumn and Spring farm sales to extend or replace our equipment. As it turned out the upgrading of our haymaking equipment year by year struggled to keep pace with development of new inventions. We were always several steps behind because we were dependent on buying second hand machines of the last stage but one. Naturally these only became available some time after the new inventions were established. For example when we were setting up most nearby farmers

HAY HARVEST

had been using Vicon Acrobats for several years to turn and rake their hay. Luckily for us these were appearing at farm sales quite frequently just when we needed one. The names of the succession of improved hay machines that we bought in the following years will have a nostalgic ring for some farmers; thus the Vicon we bought in 1973 was followed in 1974 by a Wuffler and then by a Cock-Pheasant in 1979 and that in turn by a Haybob in 1983, a very superior creature which was still serving us well when we left the farm. At sales we were always on the lookout for an extra trailer if it was not too expensive, never mind if it needed repairing. On a haytime evening when it came to beating dusk and threatening rain – as it so often did – what really helped was having enough trailers to rush the newly baled hay under cover. One more trailer stacked high if hastily was invaluable even though it creaked and wobbled. Many times we only just made it with the listing load propped upright with a row of pitch forks held by a team walking alongside, straining and praying simultaneously.

In the all-important matter of baling, our equipment was not subject to the same speed of evolution because of our youthful volunteer workers for whom the Welger baler was particularly suitable. These heavy machines had been used widely throughout the district and it was our good fortune to buy one for our first independent haytime without going further than next door: all we had to do was to trundle it up the drive through two fields and into our yard. We enjoyed the additional advantage of having expert advice close by whenever we were baffled by it. Three years later we bought a second, disused, model at a farm sale, as already described, to provide us with a complete set of spare parts for us to cannibalise as and when required, a practical insurance policy that was to prove its worth time and time again in years to come. Compared with the newer balers the Welger was slow and cumbersome but for us it had one great advantage over its faster successors; its bales of looser hay were light enough to be handled by anyone likely to want to come and help us, including young children who could easily drag or carry scattered bales to the side of a trailer where the taller and stronger helpers, wielding long pitchforks, would pass them up to those constructing the load so that it could survive the journey from field to barn. Inside the buildings, when it came to unloading the bales from the trailer onto the rising stacks, reaching up to 18 feet from the earth floor, the long forks would "pitch" the bales onto the edge of the stack to a waiting chain of eager hands, passing the bales across to the further side and packing them tight up against the roof. If every cubic foot of storage space was to be used the chain gang was gradually reduced until the last survivor was sharing his precarious perch with the final bale he was trying to squeeze in. Getting down was often an exciting challenge.

But we are running ahead of ourselves. As the hay season approached there was much to do to see that all equipment was ready for use by way of cleaning, oiling and greasing; scrubbing with wire brushes; tightening nuts and bolts; replacing missing tines; sharpening mower blades; checking the stock of binder twine. Luckily most of these were wet weather jobs which could interest and occupy the troops indoors if necessary.

Mowing marked the real beginning of haytime proper. We had bought a second-hand mower in good condition at our first sale only a month after our arrival but it had to wait almost a year before Giles was able to rehearse his only mowing lesson back at Abbotsholme and begin teaching himself to mow seriously. Of all tractor work in the hayfields he found mowing the slowest and most difficult to master. Only the few helpers who were already tractor drivers of some experience could share the many hours of mowing. It wasn't worth beginning to train those who would be staying only a short time. A level rectangle with its straight lines is relatively simple and quick to mow. Alas! few of our hayfields qualified as such. Some were bordered by the tree-lined meandering beck; two had the main track cutting across them creating triangles; in some there were patches of rushes of all shapes and sizes that had to be avoided altogether; two contained what we called "glue pots", small dangerous areas of bog where long-standing, broken-down drains had defied generations of attempted repairs, to which of course we added our own generation from time to time. The most notorious of all no-go areas for all hay machinery was "*JCB's Grave*", a big rectangular chunk of Horse Pasture jutting out into the field which we had had to fence off and plant with trees because of the treacherous quagmire that had all but swallowed up a big digger. Then again there were fields bedevilled by what we called "wilderness", strips of subsoil, rocks and tree roots thrown up by the excavation of "Offa's Dyke", the long, arterial, open gutter which all adjacent fields depended on for drainage. These strips, five or six yards wide, were waiting for the fields to be ploughed the next time they were due for reseeding, but meanwhile they diminished the area of hay very noticeably.

Long hours of mowing in field after field binds one to a farm in a way not achieved by any number of tours just inspecting and feeding sheep and cattle. You have to mow a field up and down, round and round, for its shape and undulations to sink into your understanding of its individuality, its potential and the quality of its herbage. You recognise the same patches season after season which yield richer hay than the average, and those

## HAY HARVEST

where the hay takes longer to make and the bales sometimes have to be left out overnight.

Mowing was subject to many stoppages most of which were simply frustrating technical hitches. Others however brought the kind of interest that rewards farmers for their long hours of solitude. There was the pheasant family retreating from the mower in classic fashion into the dwindling shelter of uncut hay in the middle of the field. Slicing through ten piping chicks was an unbearable option so the only alternative was adopted. This involved chasing three or four chicks at a time, now bent on individual exploration, now responding to the mother's frantic calls, and frightening them out across the open no man's land of hay already cut, to the real safety of a distant hedge and reunion with one or both parents.

Another time, mowing a heavy crop of hay some eighteen inches high, Giles was puzzled to see a portion of a narrow, well-trodden, curving track just ahead of the tractor. Inspection revealed it to form a complete circle, some seven or eight yards in diameter. The track itself was only about six inches wide. All the grasses had been beaten down flat in the same direction while on either side rose vertical walls of undisturbed grass stems. The delay before resuming mowing was stretched to allow for fetching a camera to prove it wasn't a fisherman's tale. Subsequent enquiries led us to suppose the track to have been made by hares performing a midsummer game or ritual. Though there were always hares at Bank House we never came across another such circle.

One little bit of mowing was done for us. An elderly friend of ours in the village had an orchard no tractor could reach in which she preferred to make hay rather than accumulate heaps of rotting lawn mowings. Every year she got her gardener – none other than our "adviser" with an interest in Krona – to scythe the orchard. She would tell us when it needed tedding and fetching in and we would arrive with rakes and pitch forks and a trailer. We used to load the hay onto the trailer over her garden wall to be baled back at the farm. This became a little tradition of its own, a miniature statement of our Bank House values, that provided a diversion for some of our young people and added between 16 and 30 bales a year to our store. Size and quantity are not everything!

Every haytime had its own distinct character, its own trials and tribulations and its own rewards, as our first four seasons demonstrated. Since they also marked the stages of the farm's recovery they are worth summarising:

## HAY HARVEST

**1973**      Because we expected haytime to be over before we arrived on August 13<sup>th</sup> all responsibility and equipment belonged to neighbouring contractors. We watched, learned and lent a hand.

**1974**      This was our first year standing on our own feet. We had some good helpers but no horse or fore-loader and only one tractor and two trailers. Starting from scratch we had to teach ourselves how to use each machine. The thin crop would have been even thinner if we hadn't done a huge amount of work with hand tools. We ended up on August 8th with just under 1800 bales.

**1975**      Despite poor weather which dragged out haytime until August 9th, the total hay harvest doubled to almost 3700 bales, a most encouraging reward for a full year's improvements. The heavier crop rewarded our spreading of muck and slag. Our proficiency in haymaking grew with our experience and thanks to Krona we had some of the benefits a second tractor would have provided, as for example, being able to row up and bale hay simultaneously.

The very last two trailer loads were parked for the night outside the entrance of the buildings to dry out a little more, and we went to bed. Crack! Thunder and lightening woke us at 2 a.m. and sent Giles scurrying down the track in his pyjamas to start up the tractor and hurry those trailers under cover just as the downpour began. What price experience! Our neighbour was roaring with laughter next morning. He too woke with the thunder and then heard a tractor start up. In the comfort of his bed he knew exactly what we were doing.

**1976**      This year could hardly have been more different, the year of heat-wave drought and the great moorland fire. Instead of weeks of struggle turning wet hay again and again we were exhausted by non-stop dry heat. Cracks appeared in the ground you could put your hand in. Hay had to be got in at once so it didn't shrivel up. Grass wouldn't grow again after the hay was gone and many farmers had to feed the new crop to their stock almost as soon as it was garnered, an indignity we escaped because of the remaining cover of rough growth from the years of past neglect. That year we had a second tractor and our first fore-loader and we produced over 5000 bales for the first time and all of it safely stored away by July 9th, whereas in 1978 we hadn't achieved a single bale by that date and were still rescuing hay in September.

# HAY HARVEST

There was a great deal of hand work in our early years at Bank House which employed our succession of holiday makers and other volunteers. Some of it was due to the derelict state of the farm and the necessary steps we were taking to tackle it. For instance, newly laid land drains in the worst wet patches of hayfields left ridges of soil similar to mounds in graveyards, which became booby traps for mowing, turning and baling equipment alike when hidden under the tall flowering grasses of midsummer. Most farms would have given these a very wide berth but we had willing troops and were desperately short of hay. Sometimes a gang with shovels would level the earth or even barrow some soil away. There is a limit to the risks one should take when mowing with a tractor near rocks and trees at the edges of fields especially close to those awkward strips of wilderness and in acute corners, so there is always good grass left uncut by the mower and this we attacked instead by hand with scythes, sickles and, in few cases, with suburban garden shears, and carried it off to be spread out and dried somewhere safer.

Another haytime job for everyone to join in with was pruning back hedges where they leaned out over the hay and threatened to deprive it of the wind and sun needed to dry it. Again a variety of secateurs, slashers, sickles, bush saws and the "*mighty cutter*" all came in handy. Some of what was cut off was poked back into the hedges in their thinner places. If done early enough in the season the cuttings were carted away and justified a bonfire, always a popular employment for many hands.

Weeds accounted for many of our hours of hand work. We removed some docks and thistles from hay crops before they had grown enough to be spoiled by trampling. Later, when neat rows of mown hay were turning silvery we could walk safely between them and pick out docks easily because their flowering heads turned dark red within a day or two of being cut. True, their roots remained to sprout another year but the dock seeds we took away in sackfuls to blaze with an oily crackle must have been numbered in their millions. Furthermore, had we left the dock seed heads behind in the hay, every load of muck the following year would have spread a lethal mixture of muck and dock seed, exacerbating our dock problem tenfold. Those fields that were rested and grazed allowed a more serious campaign of dock eradication. One year when we had gathered an unusually large, youthful team eager to help with the hay, disastrously prolonged rain prevented us getting in a single bale before most of them had to return home. Meanwhile, in an extraordinary array of old mackintoshes, duffle coats and anoraks, we all pulled up docks by hand, day after day. The rain had penetrated to such a depth that it was possible to pull out long, unbroken roots provided we pulled them gradually and loosened the awkward ones with a fork first. To maintain morale and interest we instituted a Mars Bar prize for the longest root pulled up in each morning and in each afternoon session. This certainly prevented thousands of docks being snapped off impatiently and left to multiply hydra-headed. The record length for a dock root was 53 cm.

Another source of handwork in haytime related to our machinery. Though it accomplished far the greater part of our work it also created some because it was sometimes unreliable being old and second-hand and driven occasionally by novices. If the knotting mechanism on the baler failed, bales were thrown out with only one string, so the foot brigade went to the rescue armed with sharp knives and spare twine. If erratic turning or gusting winds left crooked rows of hay and stray swathes which the baler was likely to miss, they were there again to straighten things out with pitchforks and long wooden rakes whose peg-like teeth if broken by rough usage could be restored by some crude dentistry on a visit to the nearest hazel in a hedge. The baler in particular needed minions in attendance, sometimes to pull bales out of its way as it circled round, sometimes to rake up hay it had failed to pick up cleanly. The longest stoppages were caused by the baler running too quickly into thick, damp hay, whereupon it choked abruptly. Then there was nothing for it but to tug out armfuls of tightly compressed hay from the bowels of the machine, handful by handful. Experienced drivers avoided this happening by spotting trouble ahead and reducing speed in time but who can always resist the temptation to hurry at the end of a long day in the field and to press on when the dew is falling and rain has been forecast?

Of all the long hours of work in haytime the most memorable and most rewarding were the afternoons and late evenings spent loading bales onto trailers and bringing them under cover. The number of trailers we possessed dictated how much unloading had to be done that night. As long as any trailer was needed back in the field to fetch in more bales it had to be unloaded at once. Each trailer's last load could be left under cover for the night and dealt with the following morning or thereafter. Until then we would continue backwards and forwards, loading and unloading. If it became clear that there were more bales than we could bring in that night the remainder had to be stood up on edge in pairs leaning against each other so that heavy dew or rain wouldn't penetrate them. Next day, assuming the sun and a breeze obliged, we had to turn them over exposing their damp inside faces to dry out before resuming the task of leading them to the buildings where we would be greeted by such a wonderful sweet scent

# HAY HARVEST

of hay it almost made us want to eat it ourselves. For all this activity the larger the team the better – what a neighbour always called our "force".

Hay spread out on the ground would get drier and lighter during the afternoons until the approach of dusk and dew began to make it heavier again. As a rule of thumb hay baled after 8pm was in danger of overheating when stacked but the work of leading bales already made used to continue long after. Many was the balmy summer night that witnessed the to and fro of tractors and trailers by headlights or moonlight: each had its own excitement. The first work brought to a halt for lack of light was unloading inside the barns where many a head bumped into roof beams before we gave in. Any expectation of rain however changed the mood from enjoyment of working to an easy rhythm to one of urgent hurry which was far more tiring.

One evening storm clouds threatened to cut our work short imminently when we still had bales for four trailer loads out in the field. There were only three of us to save them. While we were inside unloading the first we heard strange voices outside from people walking round the buildings. They came from four Dutch hikers looking for somewhere to camp. We said they were welcome but we couldn't stop even for a moment to show them where. Sizing up the situation instantly our new visitors dumped their enormous knapsacks, tent rolls included, and joined in the loading and unloading. With four fresh, burly extra helpers we were able to finish bringing in all the hay just before the storm broke. In return they enjoyed supper with us in the kitchen and a comfortable night on dry hay instead of having to pitch their tents in the dark.

On a more normal evening when the last trailer or two were being loaded in the field someone would depart to prepare supper and two others would drive off to the pub with a couple of large Victorian bedroom ewers decorated with roses to be filled with beer, a ritual that ended many a haymaking day. Publicans and their regulars alike came to recognise these ewers if not the pairs of unfamiliar haymakers who brought them, one to drive and one to prevent the beer slopping over. Moonlight made such evenings unforgettable. Nothing was more magical than watching a full moon sailing in and out of islands of small clouds, while one was reclining in comfort on top of the last wagon-load slowly swaying and jolting across the invisible fields somewhere below one towards the dark silhouette of Black Carr Buildings and the prospect of beer and supper.

Evening work in haytime

## Chapter Seventeen
## IN THE STEPS OF PIERS PLOWMAN

According to common myths the Chinese are indistinguishable from one another and only the shepherd knows individual sheep. Similarly, to the many hikers enjoying the footpaths and bridleways of Bank House our fields must have looked more or less the same as each other. When we took over the farm they all looked unappetising shades of tawny brown or russet, even grey, and it was our immediate task to make them green again. In effecting this change we soon learned how unique each field was, for each had its own character, its own geology, its own history, its own hidden drainage network and its own strategic place in the farm as a whole, all of which we had to study and respect. As early as our first year we were heartened to see our re-stocking, grazing, muck-spreading and haymaking set the best of our hayfields on their way to restoration and revived fertility. The worst, however, needed much more radical treatment and this almost always involved ploughing.

Of our fourteen possible hayfields there were only three which we were never tempted to plough, and one of these was an ancient flower meadow which has never been ploughed to this day ; all the others for various reasons, and in varying degrees, were candidates for the plough. Their two most common faults were, on the one hand, heavy infestation of weeds, especially docks, thistles, buttercups and rushes, and on the other the age, coarseness and feebleness of the existing grass ley, all of which were made much worse where there was poor drainage. Fortunately for us we were able to take advantage of generous Ministry of Agriculture subsidies which were available for sizeable drainage schemes installing deep, porous tile drains, usually requiring mechanical diggers operated by professional contractors. However, even with 70% grants these were still expensive projects for us and we had to confine them to the most urgent cases. This left us with a huge amount of hand-digging to do ourselves, opening up and rodding out shorter lengths of blocked or collapsing pipes and reconstructing old stone drains.

All drainage work spoils existing turf so it was desirable to time it to precede whatever ploughing was to be done; indeed, the two operations had to be planned together, well ahead. We had to evolve an arable strategy that would dovetail the answers to different basic questions. Which fields should we plough up? In what order? And when? What could we afford to have done for us by contractors? How much could we do effectively ourselves, considering our lack of experience and equipment and also our fluctuating and uncertain team of assistants? And what should we grow? Not surprisingly our answers to many of these questions differed from what our neighbouring farmers were doing or might have done in our place – although like the Irishman they "wouldn't have started from here". Even so, when departing from local practice we still relied on their practical advice and not only when we got into difficulties.

Agriculturally our years at Bank House belonged to the age of the "barley barons" whose fortunes were made producing huge amounts of grain to fatten livestock. Here in the North York Moors we were handicapped by both climate and terrain so that most wheat and barley grown locally compared poorly with that of farmers further south. From the beginning we could and did grow the odd field of barley but we realised that it was unlikely to bring us much profit and that we should be content if it paid for itself. Our main purpose in engaging in arable operations at all was the improvement of grassland for both pasture and hay. The best way to achieve this was the common practice of growing a cover crop of barley or some other corn and "under-sowing" it with grass seed at more or less the same time as sowing the barley so that the slower growing, shorter grasses enjoyed the shelter of the taller crop and were already established once the corn had been harvested.

Despite the large acreage in need of ploughing, we realised from the start that we should probably be limited to one big field a year, or perhaps two smaller ones, for reasons of expense, the sheer amount of work involved, and because we needed as much hay as possible without having to buy it. Every hay field ploughed meant one field less of hay for at least one season. All of which led us to think in terms of it taking us roughly ten years to get the farm back on its feet properly; and so it was to prove.

That being so there was some urgency for us to make an early start and explains why eight months before leaving Abbotsholme we spent a week in Glaisdale at the end of the Christmas holidays setting various wheels in motion. One result was that we engaged a nearby contractor to plough up four fields on which our predecessor had paid to have barley grown for him in 1972. Unfortunately he had not had them under sown with grass so they all needed ploughing again in 1973. Two of them, Middle Beck and Far Beck, were in such a poor state we were advised to leave them fallow for a further year to be ploughed and cultivated yet again in 1974: in Far Beck the barley crop had never been harvested but left to rot!

We asked our contractor to drill the best pair of fields with barley, Barley Field and Old Ley, so that when we moved in that August they were nearly ready to combine. Agreeing to sell the barley they yielded to the contractor himself was almost our only contribution to the project and provided us with our very first income from the farm of any sort.

It might have been expected that 1974 would be just a repeat of the year before as we were again to grow two fields of barley undersown with grass, but in terms of their significance to us the two seasons were as different as chalk from cheese. In the first place we were now equipped to do much of the work ourselves, using our own machines. Most exciting of all we now had a two-share plough and were champing at the bit to use it as soon as the sodden ground would allow. One morning we woke to the hard frost we had been waiting for. It was New Year's Day, symbolically the perfect date on which to be initiated into the mystery of ploughing, one of farming's archetypal activities. A sense of expectancy had been built up by our dividing the field – Middle Beck – into convenient-sized portions of "lands" with absolutely straight, parallel sides, marked out with willow rods. These we had cut from the hedge and peeled off the bark so the white cores were visible right across the field. Then using the plough itself we made a shallow groove right round the margin of the field a few yards in from the hedges and beck-side, marking out the headland for the tractor to turn on at the end of every furrow.

At last we were ready to begin. Giles lowered the plough and as the tractor moved forward the points of the ploughshares bit into the earth. At first he kept his eyes fixed on the willow rods ahead, keeping them in line so as to steer straight; then he kept twisting round to see if the shares needed adjusting. That was when a miracle was revealed. Inert, frozen clay, firm enough to build a house on, suddenly became alive, rose up writhing in two endless ribbons, turning over smoothly, unbroken, until almost upside down. Then they subsided and were just as suddenly motionless again. Only a few yards back rigid furrows stretched into the distance, seeming to say, "What life? What movement? You must have imagined it. Look, we are only dead clay. We can't move!"

That evening Mary wrote to Giles' parents:

*"We've just come in from doing our first ploughing – a beautiful dark strip against the silvery white frosted ground of Middle Beck. We're hoping that there'll be some more frosty days so we can finish it. We came home by moonlight."*

Such brief spells of poetry at Bank House punctuated longer periods of mundane reality. The weather saw to it that we were not able to finish ploughing Middle Beck until the 4th April. That enforced three months interval was probably a blessing in disguise though we hardly thought it such at the time. It forced us to concentrate our efforts on even greater responsibilities, our all-engrossing first lambing and calving seasons in particular. By the time the weather allowed us to engage in the task of converting ploughed furrows into level seed beds we had also broken new ground in another sense by acquiring and installing Jeannie, Bonus, Krona and their grunting colleagues Matilda and Placidia.

The remaining work to be done before Middle and Far Becks could yield their harvest was shared between contractors and the home team. Of the former, one spread tons of basic slag, a second drilled in the seed corn and in due course a third was to combine the ripened barley. Our contribution consisted chiefly of work suitable for youthful volunteers. Most of the tractor work was harrowing, which provided many useful hours of practice for apprentice drivers: muck-spreading being reserved for the more advanced. For everyone, there was sociable gang work: picking off stones to fill holes in the farm's tracks; trimming back hedges; cleaning out the gutters draining the fields. The big exception to gang work was the answer to the exciting challenge of sowing the grass seed. For this we reverted to an earlier technology. We bought a portable device known as a "fiddle" which could scatter seed either side of a walking operator, covering a strip of ground roughly five yards wide, pulling a rod called a "bow" backwards and forwards rhythmically, stroke for stride. It took Giles six hours of almost non-stop fiddling to cover the 4.6 acres of Middle Beck. Another apprenticeship: another skill to acquire. Then we harrowed in the grass seed very lightly so as not to disturb the barley too much and rolled the field with our own heavy roller, yet another important second hand purchase. It was May 7th.

That 1974 arable season ended in mid-September with the arrival of the combine. It was an older and smaller machine than the one used in 1973 and it belonged to a neighbour from across the beck. When it broke down, which it did every few circuits, he disappeared inside it with a heavy hammer. Loud bangs of iron on iron echoed back from the wooded hillside above. Then a succession of pieces of metal were tossed out as we watched anxiously from the sidelines. O ye of little faith! Well before sundown the combine had trundled home, leaving behind it a row of shiny plastic bags set up in the stubble, bulging with barley.

Sowing grass seed with a 'fiddle'

The four fields we had now ploughed and reseeded were encouraging proof of progress. As further evidence, we now possessed our own baler with which we baled the straw spewed out by the combine. Even more important, there were the handsome new buildings in Black Carr where we could keep it all safe and dry.

Compared with our next ploughing enterprise regenerating those first four fields turned out to have been a doddle. The largest field on the farm, previously known as "Big Field" we very soon renamed "Thistle Field" because from a distance one could see little else. Every field had its ration of thistles but none could hold a candle to the magnificent display in "Big Field" for either quality or quantity. When we embarked on its restoration we knew we were in for a battle. Long before the end it seemed much more like a whole war. Closer acquaintance revealed a subculture of docks that was remarkable considering the boulder clay masquerading as soil in which it grew. That was not all, for under this thick tangle of weeds the surface of the ground was ridged with furrows that had never been consummated with any sort of crop. Nature left to herself had had a whale of a time inviting every passing seed to try its luck and in this demonstration of the survival of the fittest grass came off very poorly.

Compounding these difficulties, which such neglect posed any hardy ploughman, were the worst drainage problems on the farm. Down each side of the field were open gutters overflowing onto the surface of the land, culminating in each case in untameable bogs, one of which we nicknamed the "Slough of Despond" after Giles had to be rescued on our first reconnaissance with the Ministry adviser. We had to fence them off permanently as we also did with the bog along the bottom of the field, the notorious "JCB's grave". As if that were not enough we found that the arterial drain running down the centre of Thistle could only cope with a trickle and so flooded unchecked all through the winter. Furthermore, there were three small "gluepots" near the bottom which every generation at Bank House had attempted to cure by adding to the complicated network of tiled drains that shifting ground below often rendered ineffective.

During our two first seasons we had let sheep and cattle find what sustenance they could in Thistle Field – one could scarcely call it grazing – but perforce always in conjunction with adjacent fields since only one of the four sides was stock proof. There was a long overgrown thorn hedge badly needing to be laid, which sheep penetrated frequently. At one time there had been a solid stone wall along the top of the field, but our predecessor had razed much of it to ground level and used the stone to fill a trench by

## IN THE STEPS OF PIERS PLOWMAN

way of making a drain. Years later a major achievement on the part of our successors was to rebuild much of that wall with stone scrounged from elsewhere round the farm. All we could hope to do was to put up wire sheep netting topped with barbed wire.

There were yet three more awkward aspects of the project in hand, which we had not encountered in the two previous years. One was that four of the five gates linking Thistle with adjacent fields needed major repair work if not total replacement: some stone gateposts had to be moved or new wooden ones installed: and all this needed doing by the time seed could be sown. All the first fields we had improved had been fairly level but Thistle lay across a considerable slope which complicated all operations especially after rain. Lastly there was the simple basic factor of size. It is a very different proposition sowing over six acres with a fiddle instead of only three or four as previously. Everything took longer to accomplish and needed longer spells of fine weather.

It will be appreciated from the above description of the state of Thistle field as we inherited it that we should need a wide range of local advice as to how we should tackle each aspect of the task we were setting ourselves. Our consultations resulted in our adopting a two stage plan. We decided to grow a catch crop of rape (brassica, not oil rape) in 1975 to provide strip feeding behind an electric fence late in that year. In 1976 after a second campaign of ploughing and cultivation we hoped to grow the usual crop of barley undersown with a permanent lay of quality grass and herbs. The simplicity of the plan didn't lure us into complacency. On the contrary we set about it with a trepidation that was soon justified. Our first attempts to plough did not augur well. Soon after starting, on May 12th, "rain stopped play" but not before the plough had run into something ominously solid. During the enforced interval we had ample time for hand digging which disclosed that covered by only six inches of soil lurked a whale-backed boulder with visible dimensions of six feet by four feet by at least two feet and still disappearing! Clearly "boulder clay" was not so called without good reason. We also had time to consider what to do with it. Sadly we passed over the idea of dynamite and settled, as every generation must have done, for leaving it alone and remembering where it was!

Slipping on wet slopes and running into boulders were comparatively minor worries. Wherever the tractor went on the unflattened ridges left by the abandoned ploughing of yesteryear, it heaved and rolled like a small boat on a choppy sea, causing the ploughshares to dig in too deeply one moment then break the surface altogether the next; leaning too much over to the left one moment then too much to the right. It was strenuous work like trying to control a bucking steed, and very slow work too. Some days no ploughing was done at all, on others very little, while occasionally long hours yielded good progress and rekindled hope. The net result was a patchwork of uneven furrows and small islands of the original, unturned weedy surface and a few bigger islands of the worst bogs which we made no attempt to plough. Since work was so slow there was little chance of finishing before bad weather returned which it did frequently, and which explains why more time was spent trying to find or mend drains than in ploughing.

If rough weather is evidence of the wrath of the gods, they surely had it in for us. On June 2$^{nd}$ we woke to driving snow! The next day the diary recorded that the ground was "the wettest since February". Even so after three and a half weeks of on and off struggle we finished ploughing as much of the field as seemed sensible to attempt and began at once to plough it all over again. This time we were able to turn over most of the patches we had missed out first time round, but not all: although we could now go much faster the second innings had its own attendant pitfalls. Because the first ploughing had broken up the surface crust there was now more danger of the tractor getting bogged down, which it did twice on successive days. On June 9$^{th}$ one neighbour generously turned out at 9.45 p.m. with his tractor to tow out ours and help extricate the plough. The next day another neighbour saw us from across the beck when we were stuck in a different bad patch and arrived unbidden with his tractor to effect a similar rescue. Then the weather showed a different face – have the gods got a wry sense of humour? – and drought developed very quickly. Henceforth our problem was how to break up clods of hard baked clay to form a decent seed bed. At least we were able to work at it all day instead of losing time because of rain.

After the second ploughing, completed in six days, we went straight on to a third but changed direction, still using the slope but diagonally so as to catch the remaining unturned areas. A new complication developed with the approach of haytime. Our only tractor was being monopolised by the ploughing but the hot weather that was baking its clods and slowing down progress there was also ripening the grass elsewhere. It became urgent to start on the hay. In the first week in July the tractor had to be withdrawn to mow several hayfields before returning to Thistle Field where a fourth and a fifth complete ploughing justified the diary's proud assertion: "Thistle now looking like a field for the first time"

Poor pasture well on the way to recovery

# IN THE STEPS OF PIERS PLOWMAN

The ploughshares were becoming less and less effective in breaking up the lumps on the surface so it was decided to hand over the task to the Triple K and other harrows. That was just in time because Giles' back began to collapse with so many days of twisting and straining. Mary and the team of volunteers assembled to make hay completed the cultivation, going over the whole field again and again. It became apparent that the simple weight of the tractor wheels was doing as much good crushing the brittle clay lumps as were the harrows being pulled. While Giles lay flat on his back in bed, the rest of the home team borrowed a spinner and spread a mixture of fertiliser and rape seed. Krona shared the job of harrowing it in with the tractor which then rolled it all, being fitted with a long wooden feeding trough adding useful extra weight and holding the stones being picked off while it went along at a slow walking pace.

After battling almost daily for two months we had sown Thistle Field and left it looking positively respectable. As if to reward us for our pains rain brought the drought to an end and helped the rape to germinate so it soon made its appearance above the surface, to our great joy. Compared to crops of rape grown elsewhere it may not have been much to shout about but remembering what had grown in Thistle Field previously we thought it wonderful, an opinion shared by our cattle in the late autumn when they had to be rationed to a strip per day. When most of the rape had been eaten the drainage contractor suddenly appeared unannounced, complete with a little hut, and embarked on a long awaited major programme of work, which included a brief spell in Thistle Field. When they cut a trench from the "gluepots" to "Offa's Dyke" they released a dark ruddy flow like lava from a volcano. They probably made ploughing easier next year, 1976, but the gluepots soon returned and remain to this day.

During the winter we concentrated on digging in gateposts and fitting them with new gates. Optimistically we began part two of our Thistle field programme as soon as possible. The diary entry in mid January was too good to be true: "G ploughed into the moonlight – ie 5.30 pm: soil much better quality than last May: much of it crumbles". Of course it got too wet again, however our contractor was able to drill it with barley on April 9th and we were ready to sow the grass seed with the fiddle next day. Tantalising breezes kept us waiting a whole week before our work was completed. In this second season barley and grass seed were sown, harrowed and rolled over a month earlier than we had even started to plough the previous year, leaving us free to start work on the next field waiting its turn for restoration.

# IN THE STEPS OF PIERS PLOWMAN

As far as Thistle was concerned our story was not quite finished yet for it takes more than a couple of seasons to eradicate the effects of a period of unbridled nature. As was to be expected the crop of barley that summer was full of weeds. In August when our neighbour from across the beck brought his little old combine it choked on the thistles and had to give up. As is usual the alternative method was slower and much more laborious, though in the long run the field benefited. We had to mow the crop with our grass cutter and leave it to dry for a few days by which time we had picked out and burned a huge amount of dock stalks with their red seed heads. Finally we were able to feed the barley and dry thistles into the combine, lead away the bagged grain, and bale the trash which the leathery mouths and tongues of our cattle were to find perfectly acceptable. For us the saga of Thistle field was nearly over. Time alone would complete the transformation. The grass ley, sown in 1976, contributed to our available grazing the following year. That September the diary actually recorded "Clover in Thistle long and luscious". In 1978 we were able to take off a light crop of hay from it and in 1979 it yielded us over a thousand bales of good quality. As swords may be beaten into ploughshares, so yesterday's battleground was reborn as one of England's pleasant pastures green, and we ourselves were a lot wiser and more experienced.

## Chapter Eighteen
## THRIFT AS A WAY OF LIFE

Life in the North York Moors was never easy, nor had the inhabitants expected it to be. Over the centuries the solutions to the practical problems facing anyone trying to wrest a modest living from this difficult northern terrain and climate had evolved into a distinctive culture which was still evident when we came to Glaisdale. These solutions, handed down from generation to generation, were shared freely with us as a folk memory, a common sense way to do things, often encapsulated in well-worn pithy sayings. Despite the revolutionary changes affecting every aspect of farming the traditional attitudes and values persisted to our considerable benefit, especially in our early days. When week after week of rain had made us doubt if it would ever be possible to sow our corn that year, it was comforting to be assured with genuine confidence, "there's always been a seed-time": whereupon we no longer felt alone with our worry but part of a traditional wisdom. It was this culture and wisdom which we had found so attractive in contrast to the popular, urban culture of the "swinging sixties", and which had turned our thoughts to farm work and brought us to Bank House.

Among the many admirable qualities and habits of our new community thrift was very prominent. People whose families had striven long and hard to achieve such prosperity as they now enjoyed knew only too well what it had cost and were not going to throw it away. They also recognised what their new neighbours were up against and were always ready to help those who helped themselves. As incomers with southern accents who had had enough money to buy a farm, albeit a derelict one, we expected and probably deserved a certain reservation and watchfulness from the locals, yet as soon as they saw how hard we worked ourselves, and that in spite of some strange ideas we came to learn and not to teach, we met with the utmost generosity and helpfulness. This would surely not have been the case if we had preached an organic gospel. That could have set up a critical barrier which had never been part of our purpose. On the rare occasions we explained our different practices, it was as a return to the traditional farming which all the older neighbours had grown up with and remembered in detail, often with a touch of nostalgia. Thrift had been a deeply engrained characteristic of traditional farming and they continued to observe it however many technical changes they were adopting the while.

# THRIFT AS A WAY OF LIFE

To us the essence of thrift was not the negative principle associated with the enforced abstemious habits of poverty as opposed to luxury and extravagance. It expressed the proper appreciation and thankfulness with regard to the true worth of things and of life itself, leading to their use with respect and care. It recognised the responsibility conferred by money and ownership not to abuse or waste whatever we have. Viewed in this way, while we were still at Abbotsholme, thrift had contributed to our increasing dislike of the wasteful habits of our ever-wealthier society, coexisting unashamedly with real poverty elsewhere. It began to affect our attitude to our privileged boarding school existence producing the sense of unease which provoked us into seeking a new way of life. It was moreover our ten years of thrift that enabled us to accumulate the savings with which we were to buy a farm, though at the time we had no particular idea in mind what purpose it would ultimately fulfil. Once we had found and bought Bank House our wider, positive understanding of thrift helped us to see ourselves as stewards rather than absolute owners. The farm together with everything that grew on it, animal or vegetable, was for our use during whatever time we should be fortunate to remain in charge. The abuse of that trust by some of those before us merely emphasised our sense of responsibility. Though we had no children of our own others would follow us who would depend on its produce for their livelihood and it was our vocation to restore the farm and leave it fit for them.

We were deliberately embarking on a life in which thrift would be a necessity as well as a choice and we were not entirely unaware of its remote relationship to the monastic vow of poverty. More to the point, our life of self-inflicted modest means linked us illogically with the great majority of humanity that had little share in affluence.

In all our early plans it was clear that we should have to make full use of every natural asset the farm possessed and rely as little as possible on any residual wealth from our previous existence to secure either necessities or comforts. Insofar as we had a coherent philosophy guiding our pattern of farming, thrift acted as a kind of cement binding many of its distinct elements together. This was most clearly so with regard to the links between organic principles, financial decisions and our attitude to the treatment of ill or wounded animals, this last being an important matter that faces all livestock farmers with difficult, sometimes heart-rending, dilemmas. The pages of earlier chapters have already given more instances than we like to remember of accidents and losses which were costly in terms of both money and extra work besides disrupting our normal programme. We never avoided summoning vets or following their advice in order to save money, but we were inclined to carry on the struggle to keep animals alive when other farmers might have cut their losses. This was not because we were more sentimental or that others were more hard-hearted, but because our situations and priorities were different. Our decisions often involved us in a great deal of nursing which might have been too expensive for other farmers but not for us with our youthful volunteers for whom nursing had special advantages. There was much for them to learn about animals, their diseases and their natural powers of recuperation, besides how to give injections, take temperatures, administer doses of medicine and so on. Even when we failed to keep a sick animal alive the experience could be valuable. On the whole if there was any uncertainty whether we should nurse an animal or have it put down, our helpers were as keen as we were to give the nursing option a chance. Similar decisions had to be made about the fostering of lambs and calves which is in effect another form of nursing. Preventing the recently bereaved foster mother from butting or kicking the unwanted fosterling away from her teats requires skill and patience and sometimes courage too. Here again we were in a position to spend more time than many other farmers and increase our chances of winning through. Our most memorable successes demonstrated the remarkable innate healing power of animals when left to themselves, also to their indomitable will to live. We related an outstanding example in a letter written while things still hung in the balance:

> "We had an animal tragedy when poor Amber, the heifer that lost her first calf at birth late in November, and whom we managed to get to foster a bought-in substitute, broke her front (lower) leg trying to cross a cattle grid to reach the said foster calf which had skipped over it unharmed. We were told by the vets nothing could be done and that she ought to be put down. So far Amber has had other ideas. She manages to get up and down and moves about on three legs and continues to suckle the calf, kept in the stable all the time and plied with lots of food and endless buckets of water. The calf is growing apace but would like a wider space to skip in. Amber's broken leg has swollen enormously and stopped waggling freely, but of course she can't put any weight on it at all. The calf is getting more of an age to survive without her if necessary."

That the story had a happy ending was due as much to the chance that the accident happened on a Saturday afternoon as to any prescience or judgement on our part. Amber had stayed on her feet till Sunday morning when we rang for the firm to come and dispatch her. They said we should have to wait through three more days and nights, which happened to be Christmas Eve, Christmas Day and Boxing Day. By the time Thursday

## THRIFT AS A WAY OF LIFE

arrived we realised that Amber wanted to live and didn't see why she wouldn't survive. The terrific swelling that developed acted as a natural splint which gradually grew strong enough to bear her weight. After six weeks, still nurturing her fosterling, we judged it safe to let her out into the yard. So Amber had her way. She lived on to rear several more calves, able to trot, to keep up with the herd though never to gallop with abandon. When we showed her to a senior vet he said he had never seen a bovine leg mend.

This experience served us well a few years later. Looking for a missing ewe, Sheba, we found her lying alone on an awkward bank with a broken leg, out of reach of any transport. She had grazed a circle round her and appeared still to have a fair appetite. So we took her food and water twice daily and otherwise left her in peace. In due course without further human intervention she was able to walk and run again, normally enough for us not to notice which she was.

Eve was the only pure Devon heifer calf surviving our first calving season to help us begin to extend our small herd, which explains our particular interest in seeing her calve safely. On the evening of April 21$^{st}$ (1976) she looked quite ready when we left her for the night with three companions. Next morning at 3 a.m. Giles paid a visit but found nothing doing. Nor again at 8 o'clock, though she was now apart from the others. By lunchtime she was trying to deliver herself. A big calf appeared but was jammed by its wide hips. First our home team, then our neighbour, and finally the vet, tried to free it. At 4 p.m. the calf was still alive but half an hour later when the vet winched it out there lay a huge, dead bull calf with its perfect russet coat. Worse still, Eve herself could neither turn to it nor stand up. The intense pressure had left her back legs paralysed. The vet explained that we must do all we could to help her to stand as soon as possible for although the nerves might recover all the time she was "down" getting no exercise, all her muscles would deteriorate. He said it was a race between nerves mending and muscles failing. He gave her a week. We left her for the night propped up with bales of straw and with hay and water close by.

Alas! In the morning she seemed little better as she lay with her fore feet stretched out, vainly pulling at them without getting any purchase by which to move the great weight of her inert body. Nine days of anxious nursing ensued. Every now and then we dragged her forward to eat fresh grass and lie on dry, unsoiled ground. Fortunately her appetite never deserted her. We milked her sufficiently to ease the pressure on her expanded udder. We removed the corpse on the fourth day hoping Eve now realised her calf was dead. On day five she managed to stand up but collapsed after fifteen seconds. We put two heifers back with her for company but when she tried to stand again one of them knocked her over, so they were banished. On the seventh morning we found her covered with thick hoar-frost which she didn't seem to notice. By the ninth day we thought we were living on borrowed time. Surely it was now or never. Giles was close beside her that afternoon when she made one of her periodic attempts to rock herself forward onto her front feet, so he added all his weight to her hind quarters. As she half rose, he grabbed the base of her tail in both hands and heaved, forwards and up. For a moment she stood there quivering, uncertain which way to fall, with Giles desperately trying to counter each lurch before it developed. After this unnatural alliance had succeeded in keeping her up for what seemed a never- ending minute, Eve suddenly took off with a bellow, going forward faster and faster with Giles hanging on for dear life so as to prevent the expected sideways collapse immediately she stopped. Round the pair went in a huge circle, coming to rest at last, still upright, where they remained for twenty minutes before Giles thought it safe to retreat and report the good news. Next morning Eve was still standing at the same spot as if too scared to lie down or to walk by herself.

By a piece of rare good fortune one of biggest cows had produced twins five weeks earlier, Holly and Ivy, and we had been looking for a foster mother. Suddenly here she was. So within a fortnight of Eve's tragedy we were at work trying to get Ivy to prefer Eve's teats to our bottles and to persuade Eve to adopt Ivy. When after a week or so we saw Eve licking Ivy we knew that thrifty Mother Nature had seen to it that every one of our cows had a calf to rear that year after all.

Sudden dramas such as we experienced with Amber, Eve and Sheba were by no means confined to livestock. The repetitive routine of farm life was interrupted surprisingly often by contrary, unexpected events. A severe drought and raging moorland fire would be succeeded by floods and landslides; a prolonged sociable glut of helpers and visitors by the silence of Christmas on our own; soaring prices would give way to plummeting markets without warning. Our ability to take them all in our stride depended on our striking the right balance between the needs of the farm itself, sufficient for its permanent health, and the needs of whatever humans could appropriately depend upon it. In supplying these human needs there was a second balance to achieve between what on the one hand we could get the farm to produce through our own efforts, and on the other the unearned bounty which nature bestowed on us freely.

## THRIFT AS A WAY OF LIFE

It might be expected that this last balance would be so uneven as to have little relevance, but reading a full year of our farm diary builds up an impressive picture of the value to us of the free gifts of nature and chance, although it couldn't be expressed in terms to satisfy an accountant. After all, what were the relative values of a winter overcoat bought as a bargain at a jumble sale and of two rabbits brought in by the cats and recorded in the said diary as "a very good supper"? However measured, nature's bounty was far from negligible and, moreover, it contributed a lot to everyone's appreciation of our way of life and was conducive to individual initiatives of all sorts. Just walking back to the farmhouse after a heavy day's work could become interesting if one was looking for tender young nettles or sorrel leaves to add to the soup, or for suitable pea sticks for the garden. In Autumn it might be hazel nuts to eat or, in Spring, a spray of catkins and a bunch of bluebells and red campion with which to civilise our concrete-floored, un-curtained kitchen, especially when guests were expected. It was always entertaining to fill one's pockets with crab apples and acorns, and listen to the unmistakable relish with which the pigs or sheep scrunched them up.

These casual, unplanned prizes were often followed up by systematic harvesting in work time, in which case instead of: *"G picked 1lb nuts while feeding pigs"*; the diary referred to: *"a rucksack of crab apples"*.

Or, if one's route home lay through the fields and hedgerows, one could be sidetracked by sloes, brambles, elderflowers or mushrooms. Usually we steered clear of other fungi; unless we were fortunate to have an expert staying with us, to lend an exotic touch to our menu. Mushrooms however were erratic in time and place. A bonanza in our second year led to our renaming "Third Field" as "Mushroom Field", a great improvement we thought, and the name has stuck to this day. That year we collected them in an open wicker laundry basket. Some went no further than the "delicious steak and mushroom supper" mentioned in the diary; most were consigned to the freezer though some were sold to the village shop at 5p. a quarter pound; others went to Whitby.

There has long been a little spinney of trees close to Old Bank House ruin which bear little plums that could be cousins to bullaces and damsons. Only the biggest and ripest are pleasant to eat raw but they stew well and make excellent chutney and jam, warranting carrying step ladders and baskets on the round trip of a mile every season. Higher up the same slope we discovered the forty - foot trunk of an ancient apple tree with a few branches, all out of reach of our ladders. Strictly speaking it was the cows

# THRIFT AS A WAY OF LIFE

that discovered it. When we moved the herd into that section of woodland one day the cows broke into a run, each hoping to be first to reach the scattered windfalls they remembered. They were right, as usual! Thereafter we used to fill our pockets with beautifully sweet eating apples before letting the herd forage there, but of course they wouldn't keep, having fallen so far.

Each summer purple bird droppings would begin to appear all over the farm. From these we learned that pigeons had decided that bilberries were ripe enough to eat and if we wanted a share we should organise an expedition. Those up our bank, almost within a stone's throw of the house, were too scanty to bother with but they told us that a few miles away we should be able to fill small biscuit tins. Those with patience did so, whereas sloes, abundant in our hedges, were ignored except by experimental connoisseurs and dabblers in gin. Brambles on the other hand, were big business every year without leaving the farm and over a longish season, which explains such diary entries as:

*"November 19<sup>th</sup> M made 24 lbs blackberry and crab-apple jam", or September 16<sup>th</sup> "M got jam factory going. 32 lbs of Old Bank House plum jam made plus 26 lbs gooseberry and crab-apple jam and also bottled 12 lbs plums."*

Not all "food for free" was fruit. Flesh too found its way onto our table, thanks largely to the invention of the motor car, but also to occasional accidents to our own animals. Though these latter cannot be accurately classified as "free" surely they do as thrift. When our Suffolk tup, Wykeham, was killed by our Kent tup, Ethelbert, Mary spent most of a day on an open trailer in the farmyard, skinning, gutting and disposing of all the parts in a race against time before the corpse got too cold. That left us with a large amount of real, strong mutton to eat and we were heartily glad when it was finished! Most genuinely "free" meat came in single meals and it was due to Mary's butchery and culinary skills that it reached our plates and was enjoyed so much. A few of our braver helpers were quite keen to learn to pluck, skin and gut hares, rabbits and birds including our own poultry. Most were just keen to eat them.

The diary was punctuated with references to our food being brought in "off the road": for example

*August 2<sup>nd</sup>* – *"dead pheasant found in hay newly died. Brought in for Mary to deal with.*

*August 18<sup>th</sup>* – *M ran over Rhode Island* – *probably ill; another chicken supper.*

# THRIFT AS A WAY OF LIFE

*September 10$^{th}$ – twice, car caught supper, returning from Folk Club: 1 grouse 2 rabbit.*

*December 11$^{th}$ – marvellous jugged hare supper (the one Krona found dead in Dock Field).*

It should be added that we never ate casualties unless they were healthy, especially the liver. We didn't keep a gun on the farm, we had enough to learn and to do without acquiring yet more skills that weren't appropriate for our young, changing household, though local friends presented us with their "bag" from time to time. Notwithstanding, such exceptional diary comments as "Jonathan caught an eel in the beck", we didn't eat from that source, not even on that occasion!

Far outweighing these relatively minor assets was our invaluable supply of firewood. We could burn as many logs as we could take the trouble to cut and fetch but, be it noted, most of the substantial, fully-grown hardwood trees were leaning out over the beck or at the far end of the farm across a steep little valley. Much of our so to speak family life centred round blazing open fires. We had had to break into a chimney in order to install an open hearth in the kitchen, and, when removing an inadequate little iron stove in the sitting room, we uncovered an old peat-burning hearth. Bank House Farm had enjoyed peat cutting rights on Egton High Moor but the practice had lapsed and the workings had been abandoned. On the Glaisdale side however the old practice survived and a few traditionalists maintain it to this day. For hot water and cooking we relied on a very old coke-burning Aga stove which the firm's own experts condemned before we moved in. They said it had a cracked cylinder which might collapse "any minute". Buying a new Aga was out of the question at the time so exercising a combination of thrift and folly we pushed our luck. By the time we left the farm we had got eighteen more years' service out of it without any expenditure – keeping our fingers crossed!

There was one gift of nature even more valuable than our unlimited wood supply and that was the spring up the bank above us. Drought and one or two temporary leaks occasionally reduced the flow from our taps to a trickle but it never failed us altogether. Without that spring there would never have been a farm just here at all. Best of all was the quality of the water. It was so pure to drink that all mains water after it tasted artificial, besides which it was beautifully soft for washing and saved us a lot of soap. When we eventually moved near to the village we used to bring back flagons of Bank House water to drink whenever possible. To prove this praise was not mere nostalgic bias we remember that our garage told us it was safe to use it to top up our car batteries instead of using bought distilled water.

Considering how many mouths were being fed, and how much and how well we all ate, it was remarkable how little actual money we spent on food, thanks to Mary's multiple skills as cook, caterer, treasurer and gardener. To appreciate the scale of her achievements one should remember not only the physical work everyone was doing but also that many helpers were still at the age of insatiable appetites which bear little apparent relation to the quantities of food consumed. Such was the prowess of our prize example that the village drama club thought it a suitable fund-raising entertainment to watch him demolish an impossibly tall pile of potatoes on their own. Bread made of stone-ground, wholemeal flour, was our other staple diet. Little white flour entered the house where it was regarded as something of a pariah. All long- term helpers learned to bake and it was intriguing that we could usually guess who had made a loaf just from its appearance despite identical materials, equipment and tuition.

In feeding us all Mary showed the same originality and energy as in all the rest of our enterprise. In all her catering thrift was ever evident and so were other guiding principles. The starting point of course was to obtain much of our food from the farm, garden and orchard. How successful the garden became is illustrated in this extract from a letter she wrote during the great drought in 1976:

> "We are eating lettuce (weighing 3 lb each), peas, sugar-peas, white turnips, spinach and Swiss chard; the beans are coming up and courgettes and marrows, so the garden is alright if we carry enough water.

This letter was updated a month later:

> "We've seven huge marrow-sized courgettes and lots of marrows coming on and scarlet runners."

The Peat Yard transformed into a vegetable garden.

By the end of October it was calculated we had also grown ten hundredweight of potatoes in the garden. Even this amount still left us needing to buy a great deal more for which we waited until the price was at its lowest and then laid in a big store topped up from time to time to last us through the rest of the year by which time we had to do a lot of de-sprouting. We followed the same policy with apples. With a cool shed we were able to store any amount. To those from our own trees we added whenever we found a cheap supply. We made the most of the harvest festival auctions and special offers and accepted windfalls which often proved to be the "good" that the proverbial ill winds blew in our direction.

Such other fruit and vegetables as we did buy were usually in season when they too were abundant and cheaper. We minded much more that our fruit had not been sprayed to preserve it than that it might be blemished or a little bruised. Mary once arrived back from Guisborough with 18 lbs of free bananas which other shoppers thought were going bad but we reckoned were beautifully ripe. From one of Mary's letters we are told about:

*"40 lbs of gorgeous ripe tomatoes which I got for 40p for the lot so we are enjoying them at every meal". Giles added a mischievous postscript – "she got them taken out of the man's waste bin!"*

Our preference for home-made food to anything processed commercially was illustrated repeatedly in the diary:

*29.10.76 - 17 lbs chutney made; second batch.*

*29.1.78 – M made last batch of marmalade; circa 23 lbs so 97 lbs in all.*

Considerations of health as well as economy led us to mix our own fresh salad dressing instead of buying sugary mayonnaise and to make big slabs of parkin rather than buy packets of biscuits. Home-made baked beans became a substantial favourite for the hungry. Opportunism and barter also characterised our catering and saved or made a little money. Once again the diary bears witness:

*Sold eggs to passing hikers.*

*Bob brought tomatoes and went away with nuts.*

*Eight pullets taken to Botton. One sold to postman en route. He ordered four more.*

*Bought 220 apple boxes total price 50p.*

Their survivors are still useful after thirty years.

## THRIFT AS A WAY OF LIFE

Another general principle we followed when appropriate was that of bulk buying. Our flour and oats came in 32 kg sacks and, like our sugar, were stored in large stoneware crocks. We bought peanut butter and marmite in 7 lb tins and ready-mixed mustard in 5 lb jars. Cornflakes came in cardboard boxes 2 feet cubed from wholesalers specialising in supplying hotels and guest houses. Two big drums we brought with us from Abbotsholme in 1973 by courtesy of the school bursar contained more floor polish and solvent powder than we were to use before we left Bank House, but then polishing floors was not high on our agenda.

There was a corner of the back yard which came to be called "Golgotha" because it was usually littered with piles of bones great and small, obtained free from the village butcher after he had spent all the time he could afford scraping off the attached sinews and bits of flesh. Besides first contributing to our soups they occupied our sheep dog, cats and free range hens for days and days and when they finally ceased to attract any attention we found that the bones would burn fiercely in our open hearths, spitting the while. The ash eventually went back on the land.

A similar procedure of trying to suck the utmost virtue out of something other people regarded as useless, was making a drink we called "hooch". This was made out of all the leftover bits of apple (peal, core, pips, bruised windfalls, the lot) and submerging them in boiling water in stoneware jars and left on the Aga top to ferment. If they looked promising after several days and started producing tiny bubbles on top, the liquid was strained off into jugs with a little sugar and the solids put in the pig bucket. We never discovered why some hooch was as delicious as any cider and some excruciating vinegar. In either case the process stimulated intense speculation and anticipation. The pigs however were absolutely constant in approval of their share.

So all-pervading was the habit of thrift that the endless list of examples must be cut short at this point after a mere passing reference to mending and recycling. Everyone contributed whether it was by darning socks and sewing on unmatched buttons or by retrieving used staples and straightening bent nails. Almost anything could be mended but we had to admit we had met our Waterloo when it came to fragmenting rusty barbed wire or to broken glass which was beyond even our devoted commitment to Araldite.

It may be apt to conclude a chapter containing such a confusion of examples of thrift in our daily life by remembering the bees. Various civilisations have regarded bees as exemplars of different virtues for human emulation, for their industry, their self-sacrificing dedication to the common good, for their unrivalled efficiency of organisation; so why not also for their thrift? One of our greatest natural assets at Bank House lay in the unbroken miles of purple heather all round us, only a small proportion of which was exploited by the fraternity of local bee-keepers. Our activity was limited by the inescapable fact that farming and bee-keeping demand maximum attention simultaneously. Thus we repeatedly failed to notice that our bees were running short of winter feed when we were preoccupied with lambing and calving. Similarly we lost swarms while we engaged in the hayfields night after night well into the dusk.

Accepting our limitations as beekeepers we adopted a policy of leaving a large proportion of the honey to the bees, and only taking it for ourselves in good years. That honey was hard won in terms of stings per jar because apart from sound gloves our protective clothing left much to be desired! Even so we were hardly ever without a good supply of Bank House honey on the dairy shelves, and our fruit, flowers and vegetables had every chance of being fertilised. As organic farmers we enjoyed one big advantage over most neighbouring bee-keepers who used to complain that their bees were suffering from the midsummer gap after spring blossom died down. Our fields meanwhile were humming with bees and sweet with the scent of clover which heavy, regular use of artificial fertiliser had largely wiped out where traditionally it had flourished on their land. Swings and roundabouts! Each to his choice; we were happy with ours.

Bee - Boles

# Chapter Nineteen
# A WORKFORCE OF VOLUNTEERS

"Did young people exist who would enjoy a period of strenuous, unpaid work in glorious country, and if so were we likely to find them?" These were questions we had asked ourselves in the train from Derby to London back in October 1972 while we dithered over making a bid for Bank House. We asked them again once the farm was ours, and from time to time during the following months of planning and preparation. Strange to say we were too busy to worry that we had made no progress on this vital aspect of our venture. Instead we were buoyed up by our conviction – for which we had no supporting evidence – that a prospect we found so thrilling and enticing was bound to appeal to others. So it came about that we still had no promise of extended help when we arrived at the farm to take up the cudgels. If ever there were cases of letting the morrow look after itself this was surely one.

After that it seemed almost miraculous to be woken at dawn on our first morning by the totally unexpected arrival of two strong ex-pupils ready to help us move in. Far from being a one-off slice of good fortune it proved to be the harbinger of a tide of support that was to characterise our life at Bank House as a whole, a tide that wasn't to turn until Christmas and then only to flow in again with the New Year once we had enjoyed our first rest.

The duplicated letter we sent out that year instead of Christmas cards was by way of a report on our first four months of farming. It concluded by saying: *"our visitors' book has fifty-one names in it and we hope yours will be added before long – but don't forget to bring your gumboots!"* This letter and its annual successors played an important role in our story. Besides keeping friends and relations abreast of what was happening at the farm they provided an opportunity for us to explain what we were trying to do. They also conveyed the flavour of life here and were handed round to anyone who showed an interest, and so became recruiting material almost by accident. Naturally people who stayed and worked with us were added to the list of recipients which after a while grew to contain some two hundred names. The following paragraphs from the letter for Christmas 1976 shows the system in operation:

*"When we decided to launch out from the homely harbour of Abbotsholme and the familiar habits of the teaching world we knew very little of what lay ahead. We didn't know quite what we were looking for except that it should be a place where we could live the good life and share it with others,*

*not a self-sufficient utopia apart from the suffering of a collapsing civilisation, but sufficiently independent for us to live and work by criteria of our own choice, yet still involved with the parent society and able to make a small contribution towards curing its malaise and assisting its victims. A place to help others – what presumption! Only now do we begin to realise that the reverse is just as true and more remarkable.*

*Our debt to you all is incalculable and far greater than you may have realised. Not only do you come and share our good life, but you create it. Without you it would not exist. You come with eager energy and spur us on to major projects beyond our routine resources. Through your fresh eyes we see again the beauty and the purpose that familiarity might otherwise dull. Our debt to our more muscular guests is too obvious to need explanation, but there may be self-confessed drones among you who fail to recognise the value of your role. We relish your infectious holiday mood; you keep us in touch with the rich variety of society at large, so do not stay away in modesty.*

*Our visitors' book records that in 1976 we have enjoyed the company worthy of the Canterbury Pilgrims, including a ballet dancer, flautist, GP, teachers, artists, civil servants, a conductor, psychiatrist, gardener, hospital patient, the head of a school for maladjusted boys, agricultural advisor, surgeon, monk, college domestic and college principal, computer lecturer, gentleman's gentleman, housewives, Rhodesian farmer, Parliamentary candidate, craftswoman, business executive, Reuters' trainee, students galore from school to post-graduate and the fancy-free in search of the Philosopher's Stone. How could life be cloistered or dull with such itinerant company and an ever-widening circle of good neighbours ?"*

This splendid stream of visitors enriched life at Bank House immensely and restored morale if ever it was threatened by the adversities of farming. Most who signed the visitors' book regarded time spent with us as a novel sort of holiday, happy to offer their services freely even though at times they did as much work as any of us. It is impossible to evaluate fairly the accumulative effect of the help all these visitors gave us. Without them we might well have lost heart when the going was rough, yet it is no disrespect to them to distinguish between them and our long-term helpers without whom we should not have been able to manage at all.

Krona, her daughter and visitors at Arnecliff End

# A WORKFORCE OF VOLUNTEERS

For want of a better term we have used the word "helper" throughout this book, as we did at the time, to denote volunteers who came specifically to join in our working routine full time whether for a week or two or for much longer. They provided the backbone of our workforce which other visitors embellished. While they were with us they were denying themselves the chance of paid employment elsewhere, a sacrifice we were not in a position to compensate in those early years before the farm began to make even a modest profit. All we could do was provide them with full keep on a family basis and offer them a meagre sum of pocket money which some very generously forewent. This arrangement was not some far-sighted plan on our part but a reaction to situations as they arose.

Our first, pioneering helper started with us when we were completing our first fortnight. He was the only one of the fifty-one visitors during the remainder of 1973 who stayed more than a week. He helped us inaugurate our reign at Bank House. Though only sixteen he was very strong and capable of a man's work. Before coming to us he had already done far more tractor work than Giles and this skill and experience was a real boon to us and established an element of partnership in our relationship. It was unfortunate for him that he lacked company of his own age during most of his time here, whereas later young helpers usually turned up in twos or threes, so he left us a week before Christmas proudly taking home with him a rabbit hutch he had made in the evenings. It was of particular encouragement to us that he was the first to support our hopes that our Bank House regime could help young people through a difficult patch after they had failed to find fulfilment and self-confidence in their school days. He was not to be the only one.

The New Year was to teach us a great deal about our reliance on helpers and how to make our partnership fruitful for all concerned. The longer they stayed, gaining experience and skill, the more they could be left to themselves without supervision. The time of year of their stay was as important as its length. The lambing season and hay time for example needed large teams and as did the days of sheep clipping and dipping. In contrast there were stretches of late autumn and winter when we could cope well enough with only one assistant or even without that. In hindsight it is quite remarkable that we didn't suffer more from such a haphazard system – if system it can be called – though there were anxious times when we wondered where badly needed help would come from. As it happened we were never stranded for long. Kind friends came to our rescue when it was most needed, as when Giles had an accident on a beam just before a Christmas which we were expecting to spend alone.

## A WORKFORCE OF VOLUNTEERS

Who were these hardy, trusting and in some cases, needy people? And what had we to offer? One of the few generalisations about them that might stand up to scrutiny is that their lives were not steaming smoothly along predestined rails in a conventional pattern. For some reason or other they wanted, or found it necessary, to spend time very differently from the way they had been, or would have been, doing. For some there was an element of a lure of adventure; for others a desire to exchange their urban surroundings for the wildness and beauty of this scenery; and for many the attraction of working with animals. In the case of older enquirers what often appealed was the escape from predominantly sedentary and intellectual occupations in order to lead a more physical life. This applied particularly to those who had suffered the pressure of exams at school without satisfaction or success. How could it have been otherwise for the lad who had attended nine different schools before joining us? It even applied when the A levels required for college entry had been secured but without developing the self-confidence they warranted.

Nearly all our helpers and visitors from a distance came to us via a chain of personal contacts, often originating from our family relations or at Abbotsholme: both sources proved particularly fruitful. Sixteen of those first fifty-one names in our visitors book were those of ex-pupils or ex-colleagues and another twelve those of our parents, aunts, nephews and nieces or cousins, full of curiosity and eager for a taste of our new life. All of them became our ambassadors. From their reports news of our need for volunteers, our open hospitality and the beauty of the place, radiated out to be reinforced by our Christmas letters. Anyone showing an interest in coming for quite a while was invited for a trial working visit. Four or five days was usually enough for them to find out if their initial, sometimes romantic, attraction could stand up to the reality of tiring, dirty work. What really mattered was that having seen the place and work, and having put real faces to the people they would be living with, they still wanted to come. If so, we were nearly always ready to give it a go.

Something we had not foreseen was that we had stumbled on a niche market for agricultural students who were usually told to get a year's practical farm experience before starting their courses. Some would-be students found this hard to procure as many farmers suspected such students might be more trouble than they were worth: training beginners for a year only to lose them when they went off for their three years at college didn't appeal to them.

## A WORKFORCE OF VOLUNTEERS

Fortunate students landed on their feet with helpful farmers who taught them a variety of skills; the less fortunate sometimes got stuck in a single groove such as milking twice daily for months on end for bosses who wanted their money's worth. One attraction of Bank House was the variety of mixed farming.

We didn't advertise ourselves but some place seekers did and we became experienced in detecting those who might fit in well with our unusual set-up. Generally we preferred to rely on personal contacts. Nevertheless some of our best helpers had been at a loose end either before or after going to university and Bank House seemed to be the solution to the problem what to do with what is now called the "gap year" or part of it. On the other hand some didn't know what to do with their degree when they had got it, like the man with a degree in accountancy who left university sure of one thing, that he didn't want to be an accountant. Someone else felt the same about engineering. For both, Bank House offered an escape for the time being.

In addition to these there were those who used part of their long school or university holidays to our mutual benefit. Giles' parents in Welwyn Garden City had been responsible for supplying us with our first helper, the lad who did so much to get us launched in our first months, and they continued to recruit for us successfully. Over coffee after church one Sunday Giles' father told the vicar's daughter that she would enjoy haymaking at Bank House! She did; many times. When a job put an end to her long holidays she sent her future husband instead. He was a medical student and when his long years of study ran out his younger brother began to turn up. Furthermore her brother once came for a fortnight.

Giles' sister and her whole family were also a fount of support through those early years. All of them beavered away on short visits even before we had moved to Glaisdale as well as after. The oldest son, himself at university, came briefly with his girlfriend. Later he sent a fellow squatter in London who was wondering whether to abandon the training he had begun. Whether or not it was a compliment to Bank House, after three weeks working with us he seemed happy to continue his training. The second son brought his best friend for two or three months after they had got their degrees whereas, the third brother came for the first three months of his gap year. He hadn't settled in long when he was pursued by a sequence of three girls who had been at school with him. Later, one of the three girls, then at York University, sent a fellow student in urgent need of a short retreat which proved so restorative he returned for a full year after finishing his degree course.

## A WORKFORCE OF VOLUNTEERS

The only sister of these three nephews didn't make an extended stay but made up for it in short visits in eight different years, in one of which she was proud to have learned how to break up large boulders with a sledge hammer and wedges. Two girls she had been at school with were among our valiant helpers: one for a whole year, which included enduring our most arctic winter without ceasing to smile. Another contact of that family came for most of a year after buying himself out of the army which he had joined on the sudden death of his father. He returned two years later for a further thirty-two weeks, with much increased pocket money because of his extra experience.

Two helpers who stayed longer than most came to us through our connection with Botton Village. One, by the arrangement involving Jeannie and a load of old hay, was Dick Bishop about whom more anon. Sally, the other we met when we were attending a short agricultural conference at Botton and learned that she was looking for experience with working horses. She figures prominently in the chapter about Krona. A letter written soon after her arrival included the comments: "She won't settle here without in some ways altering the place, which is not necessarily a bad thing but she's worth any derangement so far and we very much hope she'll want to stay a year or more." She did.

These examples illustrate the variety of contacts that brought us our helpers; they also reveal the major role of chance in our assembling a workforce. Any day the post or a phone call might put us in touch with a total stranger who in a matter of a few days would be embarking on a stay with us of anything from a weekend to a year; even, in one case, for four years. Often of course such calls came to nothing as with the call from a friend out of Mary's wartime past in the *"Wrens"*, whose enquiry is recorded in the diary's short hand: "Can we have three Italians mid June – mid July son of Godfather's friend + two friends, all aged 15?" We no longer remember why they didn't come but we cannot forget another phone call the same month from a very French voice in very broken English. Could Monsieur Noisy – the speaker – come and work with us? Giles began to explain the many reasons why it would not be possible but was so charmed by the invisible applicant that before he put the phone down he had agreed to a visit "starting next Monday"! Nothing venture, nothing gain.

In spite of his suffering horribly from hay fever and having to keep his distance from most haymaking activity, Daniel Noisy proved such an asset that the ten days turned into eighteen, after which, at our request, he returned for a further two months. He took over care of the hens, collected all the eggs and got to know not only which hen had laid which egg, but also which nettles he was likely to find it among. He made wonderfully varied salads and taught Mary a simple, tasty pasta recipe which remains our standard Sunday lunch to this day. Alas the few French folk songs he taught us have not lasted so well. We shall never know what it cost him but he did help unload lots of baled hay when we were hard pressed. So the lack of recommendation from someone who knew us did not mean that an applicant was unsuitable, nor did a gilt edged introduction guarantee success, as was demonstrated by the case of a lad we described in a letter to Giles' parents who were always anxious to know if we had enough responsible help. It is quoted at length because of the light it throws on the predicament of a great many young people in the late 1970's.

> *"Our latest inmate is not likely to stay, not because it wouldn't do him any good but because he isn't interested. Another casualty of the educational and political times – breaking off school in his second year in the 6$^{th}$, abandoning A-level without any serious alternative. A casualty too of a divorce – so it's difficult to know which trouble to help in what way. He's in a general confusion which includes going on anti-nuclear marches, wanting to be a helper of mentally ill people since a friend of his has broken down, and turning vegetarian. Oh dear! How simple it must be to be the son of a materialistic miner who just wants more money."*

Though he had been planning a long stay he passed out of our lives after only two weeks.

A friend of ours in the South, coming across a Chinese lad from Hong Kong studying Law in London, took pity on him for having to spend his holidays washing up in a huge Butlin's camp. Could we do better for him? When he got here however the tables were soon turned. He was troubled to see the lady of the house doing farm work and would rush to carry things for her. He pitied us for our isolation from civilization. How could we bear living in such empty space? And the silence! Ten days of country living was as much as he could stand and he scurried back eagerly to the comforting crowds, noise and steam of the Butlin's kitchen.

Another misfit at Bank House was introduced by a pantomime friend of ours in the village who was, more relevantly, a down to earth social worker with a heart of gold. She was trying to place a sixteen year old boy from the Shetlands, a case of ill-matched fostering. Because of the distance we agreed to a six week trial instead of the usual one, but it proved more than enough. His quick intelligence was harnessed only to doing what he wanted and avoiding all else and he made the most of occasionally genuine hay fever.

We couldn't trust him with vulnerable animals (the cats moved out of the kitchen); he was already hooked on cigarettes; the evidence of his aeromodelling balsa knife remains today as a chunk out of our dining room table.

Such disappointments however were far outweighed by our slices of good fortune as with the unheralded appearance of Deanne whom we had not even heard of when she rode into our farmyard on her motorbike. She had met both Sally and Giles' sister at a conference in Devon and decided on the spur of the moment to try her luck here while she was waiting to go to Tanzania. She stayed with us for six weeks during which her unfailing energy and intelligent interest were a great boon especially since we were busy lambing much of the time. Some you win! Some you don't!

To understand our position with regard to our helpers some unusual aspects need to be remembered. One is that they were neither employed nor paid but were volunteers. Another is that we didn't choose them, they found and chose us. Our role consisted of saying "yes" or "no" to their approaches. Furthermore the adage that 'beggars can't be choosers' applied most of the time so a high proportion of applicants were accepted. In any case it was one of our basic purposes at Bank House that we were here for their sakes at least as much as their coming might be for ours. In considering any suggestion that a person should join us we were weighing up the chances of their time here serving their needs as well as ours, as with the two lads who attended day release courses at Guisborough while they were with us. We had an instinctive hunch that these contrary interests would somehow balance out in the long run and by and large they did, though not without leaving us to get through some testing times.

As will be evident already there were a few people whose positive contribution on the farm provided meagre recompense for the trouble their presence involved. Far more often however we felt almost guilty that instead of fulfilling our original intention of helping the young it was we who were the chief beneficiaries. In this game we met with distinctly fewer snakes than ladders and fortunately the ladders were usually the longer.

It was probably to be expected that young people idealistic enough to opt for the limitations and hardships of our Bank House life would have the imagination and compassion to be good companions for the less able and more needy. We witnessed this relationship in our first winter with Dick Bishop and Gerard. Dick was an inarticulate probationer suffering differing degrees of physical, educational and emotional handicap – in rising scale. Gerard, born to be a scholar, was wise and charitable. For three months,

A WORKFORCE OF VOLUNTEERS

which included our calamitous first calving season, they were our only resident helpers. At a time when cash flow was an acute problem Dick was adamant that he would not wear one of the long, heavy civil service greatcoats that we had been lucky to buy very cheaply for each of us. Instead his heart was set on a more expensive, skimpy donkey jacket with a collar of rabbit fur genuine enough for our sheep dog to attack and chew it up.

All the time Dick had been at Botton the community had received payment from, oddly enough, the Ministry of Labour, in recognition of his handicaps. When Botton asked us to take him on they assumed we should receive the same funding but thanks to incomprehensible bureaucracy it was discontinued, leaving us with months of negotiation we could well have done without, and finally an unsatisfactory settlement.

As it turned out Dick was to live with us for nearly two and a half years, overlapping with many different helpers and short term visitors all of whom recognised his special needs and assisted in meeting them. If anyone, it was we who were thought at times to be unduly impatient and critical, but then it fell to us to cope with his unhygienic habits, the bedding damaged by his cigarettes, his helping himself to money from Mary's handbag and his getting into debt to a mail order company when he couldn't resist the temptation posed by their advertisement for a football which was delivered to our door, though he disclaimed knowing about it. On the other hand, Dick's physical strength was useful provided he was working in company. It was a bit of a give- away that he always knew whether the post van or an expected load of straw was passing along the road nearly a mile away on the other side of the dale, or if the neighbour's car he had seen disappear towards the village had yet come back again. His laziness and uncleanliness we could surely have put up with indefinitely, however we had to send him packing when he threw a brick at Mary in a temper. He had always been liable to blow a fuse when criticized. In the culture of his upbringing men told women what to do and it wasn't manly to accept the opposite.

Only one helper stayed with us longer than Dick and that was Carl whose only similarity to Dick was that for both life was permanently askew because of the lack of a stable loving home life throughout childhood. Carl had been placed at the boarding school for boys in need where Giles had taught for six years, long before. An emotional crisis led him to clear out of his foster home and abandon a suitable training course. He was on his beam ends near the school, sharing crisps and condensed milk with his dog and in danger of involvement with bad company. The farmer's wife at the school, to whom he had turned and who had read our Christmas letters each year, appealed to us. Could we take him? And his dog, his only reliable source of love? "Yes", we said, "but he may have to choose between his dog and us if it cannot be controlled!" He arrived, brought by a County social worker, with the dog and also (horrors!) his cat, which gave all our cats flu! The dog proved as amiable and charming as its owner and both survived here on and off for nearly four years. He was very rewarding to live with, being perky, imaginative and full of enterprise, teaching himself about motorbikes after naively buying a poor one, trying to cross a deep ford on it and being swept away. He pushed it home – mostly uphill – dried it out and spent months teaching himself to restore it. His life with us and apparently since is like a soap opera of unlikely episodes which frequently ended with his running away. It upset him when the others went on holiday, first to go home and then to university. He also got entangled with the charms of a girl who "helped" us here for two weeks or so. Once he set out hitching with his dog, hoping to find paid employment. It didn't work out. After four painful months with the thought of winter to come, he wrote out of the blue pleading to come back. For his last year with us he became our lodger contributing to his keep with some earnings from a series of local jobs which we helped to find for him. It was an important step forward. He persisted doggedly in trying to find his real mother and eventually succeeded so he left us to go and live with her and three younger siblings. "Happy ending" he wrote in our visitors' book with characteristic optimism and inaccuracy, giving the wrong year.

It was in Carl's time that our regime of volunteer resident helpers came into its own and life in Bank House was almost continuously sociable as well as busy. The farm was finding its feet with its new buildings in Black Carr for the cattle and the beginning of improvements to the house for us people. When we moved in, besides our own bedroom and a fair-sized guest room, there were only two very small single bedrooms until we had the house re-roofed in our first Spring. We took advantage of that operation to insert three dormer windows into the roofing of the dark attic which ran from end to end of the house. Originally it was used for the storage of corn, sacks of which were hauled up from the front hall through two successive trap doors by means of a wooden winch which is still in position. Our improvement created two good rooms for our expanding household; one became a boys' dormitory with three beds and the other we called our "library" having a large trestle table surrounded by lots of Heath Robinson bookshelves consisting of loose planks held up with builders' bricks. This gave all of us more space to do our own thing provided we could stand the cold, for in the absence of central heating the hardy winter reader relied on the fact that hot air rises – from two stories below! There was usually a sleeping sack available.

In terms of human comfort Bank House was pretty spare; undecorated too until the arrival of a long-term helper who came on the understanding that she didn't have anything to do with the animals! She said she would decorate the rooms instead and made a very good start, much of it when the rest of us were all out of the way, tucked up in bed. Of course the animals won, so she ended up doing both.

From the diary one gets a good picture of our "family" domestic evenings: for example,

*"S to Young Farmers, G mended jersey, O mended coat and practised song, T wrote letters, D drew! M cooked."* Another evening is recorded: *"M to orchestra in Whitby, G & N & C to rehearsal in village, Gus had peaceful evening phoned home, read books, put on records."*

Naturally we often had to entertain ourselves. The reception of our black and white television in the sitting room suffered from our surrounding moors and sudden valleys so the set wasn't used as much as might have been expected. Instead it was often out-rivalled by our collection of games in the kitchen. There were dominoes, drafts, chess, bagatelle, tiddly-winks, spinning tops, nine men's' Morris, shove-halfpenny and spillikins not to mention Scrabble and all sorts of different card games. Crosswords tended to be communal events but three dimensional wooden puzzles were definitely individual struggles as of course was solitaire. Usually our team boasted a guitar player, sometimes two, subject to an entirely voluntary in-house tradition that playing a guitar involved sitting on one's bed.

We were proud possessors of a quality croquet set which we had played with often while at Abbotsholme and we persisted in the fond hope of playing at Bank House, despite the absence of anything better than a newly cleared hayfield to play on. The only serious attempt we made was hardly croquet, less because of the rough ground than because of the playful alliance of kindred spirits, Wilf and Diligence, puppy and foal, who competed with each other in chasing every ball that moved. We were much more successful playing boules in the yard where the steel balls bounced sharply off the stones.

Our slowness to find and join in with leisure activities away from the farm was only partly due to the extent to which our time and energy were exhausted by the demands of rescuing house and farm from dereliction. Even more responsible was the lack of transport. No bus service called at

Glaisdale and there were only four trains a day in each direction from our station and none on Sundays in winter, nor was there ever a train late enough to bring us home from an evening in Whitby or Middlesbrough. Our one, ancient bicycle, devoid of three speed gearing, was always available provided rough treatment hadn't put it out of action and the fact been forgotten. Even at its best cycling was not exactly an attractive proposition since there was no route out of Glaisdale that wasn't beset by fierce gradients. As a result few entertainments beyond the village were accessible without the use of our car and those in the village itself often involved walking two miles each way. Even so we all began to find our way to the village for our own purposes: Dick to buy cigarettes, Carl to attend art classes, Sally for meetings of the Young Farmers' Club, Mary to WI evenings, (Giles too on "open" days) and both to attend Church services whenever farming permitted so that Mary was soon elected to the PCC. Naturally everyone patronised one or other of Glaisdale's three pubs. Another attraction from 1977 onwards was the Beggar's Bridge Players drama group which was a legacy of the Jubilee celebrations that summer for which village tradition expected a play to be specially written and produced. The brave moving spirit was soon in difficulties. Only one man in Glaisdale could be persuaded to appear on the stage until she bemoaned the fact to Mary in the butcher's shop, whereupon Mary promptly volunteered Giles as a sacrifice and Giles, following suit, volunteered our only male helper at the time to keep him company. Thus casually was born a fruitful association of the Beggar's Bridge Players and Bank House helpers whose circle of acquaintance and sociability in the village was thus permanently enhanced.

A break-through to participation in Whitby's social life came with two discoveries. The first was of the "Three Arts Club" which provided a subscription season each winter with a varied programme including recitals, plays, opera, ballet and lectures, usually put on in the Spa Theatre. There were sufficient supporters in our neighbouring villages of Glaisdale and Lealholm to warrant hiring a coach. Besides enjoying the shows we made friends with other local people as we journeyed to and fro.

The second discovery was of the Whitby Folk Club which encouraged beginners without detracting from the very high standards of performance by its regular members as well as its guest artists. The club met in a pub which also hosted meetings of an informal, early "green" discussion group at which Mary made more friends who were interested in what we were doing at Bank House. This group led some of our helpers to realise that we were part of a wider movement.

# A WORKFORCE OF VOLUNTEERS

No account of our volunteer workforce can be complete which does not cover its local element. It was only to be expected that news of the unusual goings on at Bank House should interest local people at large and attract a few who like our resident helpers had time and energy to spare. What distinguished them of course was their having homes and families nearby. We hadn't envisaged fitting such people into our routine so each case was a unique arrangement. For instance there was the poor lass who lasted only a week, living in as one of the gang. Her parents' business folded because of the miners' strike which proved disastrous for Whitby as a whole. She could hardly have been less suitable as is shown by these notes of description written at the time.

"Poor education – feels unwanted – grand ideas, not ready for hard work – expects everything on a plate – likes being here as a sociable place but didn't do much"

Another short-lived arrangement concerned a teenaged girl in Glaisdale. After her father died her mother and siblings moved away from the village but she wanted to stay to complete her exam year at Whitby school, though we suspected that a boyfriend also came into the equation. It was agreed that she should board with us during the weeks, however she was soon unwell and the doctor diagnosed that she had an irremediable allergy to cats and she had to leave.

Yet another scheme of ours came unstuck. We were acutely aware of the lack of suitable practical openings for local school-leavers but we didn't see how we could integrate them with our system of resident helpers, nor how we could take on the extra responsibility of overseeing them ourselves. The idea we came up with instead was for us to offer the free use of our nutwood for unemployed school-leavers to be trained in woodmanship; to manage the wood, coppicing a different area each year and earning an income by producing and selling firewood, hurdles, lobster pots, beanpoles, thatching stobs, pea-sticks, etc. We found a supervisor who was paid by some Government fund to launch the project. He began with the only two willing youths he could find, all others apparently preferred idleness and the dole. The scheme died painlessly when it was discovered that the more promising of the two lads was already committed to the merchant navy and was merely filling in time. The other lost all interest.

Much more in our line was the Glaisdale boy who surprised his parents by suddenly asking for gumboots for his twelfth birthday so that he could do some farmwork. He fitted in very easily on Saturday mornings tagging along with whatever we happened to be doing.

More significant than the above local contacts was our link with Bert from Whitby. One evening when we were at the Folk Club we noticed two friends totally absorbed with their matchstick game apparently oblivious of the music. Suddenly one of them came across and accosted us. Were we the organic farmers from Glaisdale that someone had told him about? We came to know that Bert was brought up farming locally and never wanted to do anything else. Tragically, arthritis cut short his father's working life and the family had to sell up and move into a housing estate in town, whereupon Bert went for two years as a VSO worker to Papua New Guinea. It was after his return that we met him and found that he was at a loose end, longing to use his energy and skills. Besides his knowledge of most aspects of farming he had a special bent for our particular weakness; all things mechanical, electrical and technical, including plumbing and metal work.

His visits to Bank House became frequent though irregular, often extending through and long after supper until he became as much a member of our household as many residents. When he was part way through a project he might come more or less daily. It was he who thought a sledge for Krona would be useful. He designed it, produced much of the material for it and did most of the work on it. He also designed and built a header tank for the water supply for our new Black Carr buildings, some fifty feet up the bank above them and followed that up with the plumbing work for the water bowls inside where he also installed some minimal electric lighting. When he saw how many disintegrating rusty iron gates there were on the farm he brought a gas cylinder and his welding kit and gave many of them a new lease of life. Everything he did was individual and cheap; if he didn't have spare parts himself he knew someone else who did. Bert had become interested in organic farming and had joined the Soil Association before we met him which helps to explain why he fitted in so easily. Most of what we were doing was similar to the farming he would have liked to be doing himself. He taught us a lot, solved many of our problems and made our lives more interesting. For instance, if we had a sick animal and were discussing how to treat it, he would in all seriousness draw on his experience with wart hogs in Papua. Into the bargain he must have saved us a lot of money.

This chapter has concentrated chiefly on our early years at Bank House. Reflecting on it as a whole it is difficult to avoid the conclusion that we were born lucky. Without any practical plan of campaign how to find or reward an essential workforce, we had embarked on a venture that could not succeed without the huge amount of unpaid labour that the rest of this book must by now have made only too plain. Our survival could hardly be

## A WORKFORCE OF VOLUNTEERS

said to justify our policy, or lack of it. A charitable verdict would simply be that we 'got away with it'. More important however is what our story proved about the sense of adventure, stamina and generosity of the young.

## Chapter Twenty
## PROGRESS AND CHANGE

In describing how we set about restoring Bank House Farm the previous chapters concentrated very naturally on our early years. This chapter is designed to satisfy the curiosity of readers who would like to follow the story a little further in various aspects covered so far. It was one of the attractions of those pioneering days when so much needed doing, that progress was evident wherever one looked, rewarding one's efforts and stimulating further activity. For instance the gangs we assembled for hay-making rejoiced to see how many more bales we made each year as the total rose from 1,798 in 1974, to 3,676 in 1975 and to 5,112 in 1976, after which it levelled off having reached the point beyond which we did not wish to increase the work or lose more grazing. It was particularly encouraging to see how much more a field produced as a result of our ploughing and reseeding. Thistle Field and Horse Pasture, our two biggest hayfields, each produced over a thousand bales in 1988. Of course quantity is not everything, nor could we claim all the credit for such bumper crops; weather played a huge part in it, but all in all we had much to be satisfied with from our efforts. Quality was a much more subjective matter when considering progress. A letter we wrote in October 1977 shows how much we were uplifted by some of the improvements:-

*"We did wish you could have been with us some two weeks back when letting the cattle into Thistle Field. Never had we had such a rich, dark green feast to offer them: a thick, healthy, fresh menu of clover and grasses with remarkably few dock leaves or thistles in sight (they'll be back next year!) – but an even carpet of food covering the whole field – and firm enough under foot. I think that has been the most satisfying and rewarding transformation we've achieved so far.*

*And you who will remember the wilderness of ridged clay and weeds, would appreciate it particularly. So it is a vision to treasure alongside the fat lambs for this year. It's nice to think that Dock Field is well on the way to provide similar nourishment next year."*

That last sentence introduced another important yardstick of progress, the proportion of our reseeding programme that was accomplished of the eight fields we had originally considered needed tackling. When we started at Bank House visitors often asked us how long we expected it would take us to get the farm back on its feet in reasonable, if not apple pie, order. The figure of ten years that we plucked out of the air as an answer proved quite

## PROGRESS AND CHANGE

near the mark, for in 1983 we completed those eight fields and were planning to begin ploughing them for a second time round.

Another measure of progress was the rising prices of the beasts we sold, though at any one sale this might reflect the vagaries of the market as a whole as much as the better quality of our produce in particular. Our first sale in 1975 was a disaster, with the highest bid for one of our beasts being only £47, which was a reflection of the poor fertility of the farm at the time. However the steady, general rise thereafter was clearly a sign of improvement. Our best yearling in 1976 fetched £206, that in 1978, £253, and £326 in 1981, while we got £402 for Polyphemus in 1988.

Our lambs improved similarly though more erratically. A letter dated September 1981 records:-

> "We had a very successful sheep sale yesterday. It was particularly gratifying not just that the prices were high – even our smaller ones fetched more than our previous record – but they were all very good, plump, healthy lambs, among the best at the sale, and all our neighbours know good animals when they see them."

Lest we paint too rosy a picture it should be acknowledged that many neighbours regularly produced bigger animals that sold for higher prices than ours. Such comparison however ignores half the story. It was never our overriding aim to breed the largest and most popular animals whatever the means available. There was no doubt had we followed common practice, particularly with regard to the use of drugs and chemicals, our crops and livestock would have grown faster and brought in greater profits. In this we suffered from being ahead of the times, for then we enjoyed no equivalent to the organic premiums that have become common since. Certainly some of our customers chose to buy from us because they liked the obvious health, placidity and flavour of our animals and felt secure knowing they had not been overdosed with antibiotics.

Our farming must have puzzled local farmers for we decided each issue as best suited our young workforce and generally followed compromise solutions as in the case of spot-spraying individual weeds rather than blanket spraying whole fields or not spraying anything. There was one matter which we settled on principle and which we never abandoned: we would not de-horn our cattle even though it narrowed the field of potential buyers and reduced what the remainder were prepared to pay.

Some of the changes we made were improvements we always intended to carry out which had had to wait their time and opportunity. One that didn't have to wait long was acquiring, and learning how to control, a sheepdog. Nothing illustrates more clearly the remarkable kindness of our fellow farmers in shepherding us through the most difficult period of our inexperience than their help over the matter of sheepdogs. "You'll be wanting a dog, then?" said one during our first week. Two more we met watching the sheepdog trials at our first Egton Show a fortnight later, advised us to start with an old dog and said they'd look out for one. Within a further fortnight we were told that a farmer up the dale had a dog to sell that would suit us because it was proving too slow for his purposes. Everything was more or less arranged for us between our neighbours so that in only our fifth week at Bank House six-year-old Border collie Kim took up guard duty in our yard. The generous arrangement was that if we still wanted him after a trial period we should pay just £5 plus a little for his collar and chain. Within six weeks Kim had become so useful we were only too pleased to complete the transaction and become the proud owners of a sheepdog. Bank House has never since been without one for any length of time.

Giles never attained great expertise in managing sheep with Kim who at six years old was rather set in his ways and probably right to question his new master's competence. This was hardly surprising since time for regular training sessions was not found often enough: nevertheless Kim's advent marked the vital step forward in stock control from two legs to four legs, as he settled happily into his role as a general farm dog with a special interest in sheep. Much of the fluctuating progress that ensued concerned widening the circle of dog handlers for whom it was a red-letter day to find Kim responding more or less correctly to their commands. Such was the situation for five years until to our dismay we noticed him beginning to bump into things or fail to see the distant sheep we wanted him to round up. By the time we wrote our Christmas letter in 1978 we were commenting on his remarkable contentment with the simplest pleasures of canine life despite his near total blindness.

That discovery led to our acquisition of Wilf whose training and subsequent career were to constitute the second stage of our sheepdog history. This time we were able to choose him ourselves, or rather he chose us. When we inspected the litter of five romping puppies one of them advanced deliberately though unsteadily towards us and introduced himself as if to say, "I'm the one you want." He certainly was. He loved all the human race but us in particular.

## PROGRESS AND CHANGE

There followed a period of partnership between the ageing Kim and the rapidly plumping, irrepressible Wilf, described by Giles in letters to his parents.

*"Wilf shows much interest in sheep – which is awkward because I can't start training him on the sheep in lambing time. I'm concentrating on getting him to lie down when told to, which he often does, but bounces straight up again, saying - 'That's done! Now what?'"*

Much of Wilf's training was due to Kim whose every move he watched, but it owed most to his own inborn instincts. Giles, himself a raw recruit, learned much from the advice of neighbours and to a sheepdog training course arranged by the Ministry of Agriculture, but far the most to Wilf himself with whom he developed a far closer understanding than he had achieved with Kim. Giles began to notice that quite often Wilf anticipated commands by a few seconds so that by the time the word was uttered the dog was already on his way. It seems that Wilf had been scrutinising the pattern of his master's body language that preceded any actual sound, and acted on what he saw without waiting to hear.

From very early days Wilf was convinced it was his inalienable duty to police all improper behaviour be it straying livestock, squabbling hens or uninvited aeroplanes. He declared personal war on anything in the air with an engine. At the first distant sound he would rush towards the approaching intruder, barking furiously, but finding that he couldn't stop it in its tracks he changed tactics and thereafter, before the plane even reached him, he would turn and bolt off instead in the direction in which he guessed it would disappear. Far from realising it was a futile exercise he was obviously pleased with himself and would return, tail wagging prodigiously. In his eyes he was 100% successful, knowing that in the whole of his ten years' policing he had prevented every plane from landing on Bank House soil. In retrospect Giles considered that training Wilf was the most deeply rewarding of all his experiences at Bank House as the tubby ball of exuberant abandon turned into the lean, purposeful creature of total concentration that every effective sheep dog needs to be. The partnership lasted almost ten years, in fact until after Mary and Giles moved out of the farmhouse into the village, leaving Wilf at Bank House. By that time he too was considerably blind and we realised he would be happier to remain on familiar ground: in any case Giles was to continue spending all working hours at the farm with Wilf.

In those days sheepdogs were very prone to blindness. Steps taken to avoid the hereditary weakness being perpetuated had not been established long or widely enough to prevent our suffering with both Kim and Wilf. In each case the dog continued to accompany us in all our work round the farm long after his eyes were almost useless. However, noses and ears – and probably sixth senses too – proved surprisingly adequate substitutes. Naturally they were increasingly bored not seeing whatever work we were doing and took to wandering off ever more frequently, which worried us until we found that they always turned up safely back in the kennel in the farmyard. On occasion, suddenly noticing their absence, we would spot them in the distance, about to blunder into a field of cattle and we would watch anxiously as they successfully navigated their way through or round these invisible hazards.

It was remarkable how useful they remained in their work, for the sheep seemed to have no idea of the dogs' blindness and never exploited it. Luckily by the time each dog's sight was fading he had built up an extremely accurate map of the farm in his head by which to avoid circling uselessly or straying off the farm altogether. Their detailed memory knew the nearest useful gap in a hedge by which to circumnavigate a closed gate found to be barring their way. We would watch them rooting about on the trail of different scents, tails wagging now faster now slower, as they neared or lost their scented quarry of extra food such as sheep's afterbirths, stray eggs and left-open food bins.

For each the sad day came when his confidence showed signs of waning and an element of fearfulness persuaded us to call in the vet to end his darkness. Each in turn, Kim in 1980, Wilf not till after we had left the farm, died peacefully in Giles' arms after a single injection and was buried a little way up the bank where in turn they had so often watched and waited for us to emerge from our meals to summon them to the exercise of their craft which always gladdened their hearts.

Complimentary to the enormous saving of human energy which Kim and Wilf represented were continual improvements to our fences, walls and hedges, the absolute priority of which occupied a great proportion of our first three years at Bank House. Most of it took the form of wire netting surmounted by horrible barbed wire for reasons of speed and economy. All we had been able to do at the far end of the farm had been to seal off twelve acres in order to prevent our animals from tumbling down the ravine into the beck, wandering off unhindered to Beggar's Bridge en route to Whitby or spoiling the idyllic spread of bluebells and wild daffodils. Now, in what a communist government would have called our second three year plan, we were able to start rescuing those abandoned acres. The project involved

## PROGRESS AND CHANGE

almost half a mile of boundary fencing along with internal divisions in very awkward terrain. For all of this we spurned barbed wire, resorting instead to twelve-foot wooden rails in order to facilitate enjoyment of the beauty spot, ironically, just as the public bridle way status of the route through it was being allowed to lapse by the powers that be. On and off the project took over three years which included the hardest winters and deepest snows we ever encountered in Glaisdale, hence our memories of retrieving tools, posts and rails that when dropped or nudged inadvertently, slithered away down the steepest slopes like unmanned toboggans.

After that we enjoyed extending the variety of work in enclosing our livestock elsewhere. Overgrown hedges were rejuvenated by the time-honoured rural skill of laying where the existing hedges had enough vitality; otherwise we used cheap off-cuts from timber yards to reinforce weak spots in hedges or to block up unauthorised holes made by animals. Everyone joined in making bonfires of the huge amount of left-over branches and foliage. Dry stone walling, mostly on higher ground or round the farm buildings, was another rural craft practised with varying degrees of skill by almost everyone working on the farm. We began our apprenticeship by repairing the simpler ravages of time, the damage done by adventurous sheep or occasionally by careless hikers. For those of us with natural aptitude practice would develop real skill which sometimes found scope on a larger section of wall as when we decided to re-route the track leading up to the Dutch barn to facilitate the delivery of towering lorry loads of bought-in straw. This involved removing existing walling and rebuilding on either side of the gate and gateposts we inserted. On another occasion a cavalier driver contracted to spread tons of lime and slag on the top fields took it on himself to flatten seven yards of the sound boundary wall – out of sight of the farmyard and house, of course – and drove his load straight over it, because, he said, he couldn't get his lorry in otherwise. His promise to rebuild the wall himself never materialised, which was hardly surprising. Several years later it gave Giles an excuse to try his hand at his most substantial project of dry stone walling which he would not have dared to undertake if he had not by then attended a three-day course of tuition from the best practitioner in the district. After over thirty years it looks sounder than much of the wall on either side that wasn't knocked down.

The period of our second three year plan saw us engaged in another attractive scheme which we found too tempting to resist. The moorland skyline above Bank House was dominated by stark, rectangular pine plantations which seemed wholly out of place in our rounded moorland landscape, so we were delighted to discover that the North York Moors

National Park (hereinafter "the Park") was promoting a scheme to encourage landowners to plant a great variety of hardwood trees which it provided free, along with fencing materials and tree guards, for us to plant and protect. Bank House met the criteria for successful applications to perfection. The justification for the Park's financing private trees was their control of the choice and positions of the trees to maximise public pleasure, for not only was the whole farm open to view from moor and dale but it was riddled with rights of way.

We proposed eight small groups of trees in strategic places. Some made use of bogs or awkward corners of fields were ploughs and haymaking machines could not operate easily. Others were sited on the crest of the bank so as to break up the straight horizons of pine plantations. Our application was accepted and the work was spread out over the years 1977 – 1980. As a result we planted some two hundred trees including at least thirteen different types, from oak to holly and field maple, and from lime to rowan and cherry. The visual benefit was far greater than this small number of trees would suggest.

One of the benefits of our participation in the tree planting scheme was to introduce us to various Park officials and vice versa. It was the Park's purpose and duty to accommodate three potentially conflicting interests: the farming community; public recreation; the landscape and its wildlife. With such diverse aims it was natural for it to support changes in farming that were alternative to maximising production, consequently it was only a short step for the Park to regard our regime at Bank House as a natural ally once they got to know us. Its *Farm Scheme* comprised a series of unique, tailor-made contracts between the Park and individual farmers who undertook to avoid ecologically harmful practices in general and to carry out specific beneficial works in return for moderate agreed grants. Before launching the scheme more widely in 1990, trial runs were arranged with half a dozen suitable farms and we had the distinction of being selected as one of them. By our first contract in 1988 we agreed to a three year programme of work which included small stretches of hedge-laying and planting as well as a few days of coppicing hazels in the Nutwood.

While we were busy planting trees we were contemplating a change in our arable policy. With the first five fields that we ploughed and re-seeded we had stayed with the prevailing fashion, growing and combining barley. Then we began to realise that what was best for most farmers was not necessarily best for us in our unusual situation. Growing barley in this district was likely to be profitable if the grower used it to fatten his own

## PROGRESS AND CHANGE

beasts or if the barley was of good enough quality to sell to other farmers to finish theirs. Our main income however, came from selling off our calves as yearlings. To keep them on in significant numbers and finish them ourselves would have needed more land and more accommodation. In any case, with Bank House as it was, we were unlikely to produce good enough barley to sell profitably. Besides this, Mary's previous arable experience had been growing oats in Sussex using horses. In the past when every farm depended on horses oats were widely grown in these moorland dales too, but by the time we came to Glaisdale working horses had almost entirely been replaced by tractors, and oats were being replaced by barley. Now we had Krona, and Mary remembered that the farmers in Exmoor from whom we bought our Devon heifers had agreed with her that the best food for them was oats. Couldn't we at last do this and grow oats instead of some of the barley?

After our two-year struggle in Thistle field the next most important field to plough up had to be Dock field. On our first visit to the farm we had been puzzled to see from across the dale what we took to be a three acre square of rust red bracken. Alas we were looking at billions of dock seeds. Being the nearest field to the house and farmyard it was, acre for acre, the most valuable on the farm, of the utmost convenience for house cows and young lambs on their first days out. It was to be the scene of our dock-pulling competition for Mars bars in the wettest of hay-times and also of our first venture with oats.

As with Thistle field we planned to plough it two years running to achieve the most effective weed clearance: the first season of rape was to be followed in the second by the best possible grass lay, undersown with oats. So far so good. Then the wrong sequence of weather intervened preventing us harvesting the oats until the grass and weeds were thick amongst the oat stalks, making it unlikely that they would ever dry out so late in the season. Botton came to our rescue and lent us some 48 tripods, devices developed in lands such as Norway and Scotland that often lack haymaking weather. Each tripod of eight-foot poles wired together at the top had wire hoops holding them from slipping outwards.

Making oat hay on tripods

## PROGRESS AND CHANGE

The hoops were as it were "woven" with thick swathes of oat hay, round and round from the lowest hoop upwards until, with most of the actual tripod disappearing from sight, it was topped up using pitchforks with yet more hay forming a dome. The globe thus constructed with all its outside grass and oat stalks sloping downwards, was in effect perfectly thatched and could withstand any amount of direct rain, for weeks if necessary, while everything inside remained dry as it matured into scented hay. How much wind these tripods could stand depended partly on the skill of each builder, and friendly rivalry ensured that after a night of equinoctial gales we would rush to the bathroom window to check whose tripods had collapsed and needed putting up again.

A big advantage of the tripod system was that we only needed one or two dry days to make hay; the day the globes were built and the day they were dismantled, aired and taken into the buildings where they could be baled at leisure. The relish with which the cattle ate that crop in midwinter made all our extra labour worthwhile. We knew that we might not always be blessed with an adequate workforce and that future oat harvesting would have to be easier, especially since Horse Pasture, the next field on our list, was over twice the size of Dock field.

When discussing what to do with Horse Pasture with our neighbour, Fred, he surprised us by revealing that he had an old reaper and binder machine at the back of one of his sheds with which he used to harvest his oats, but it hadn't seen daylight for seventeen years. The more he thought about it the keener he became to try it out again. Mary caught his enthusiasm. It would be a big gamble to entrust the whole of our year's corn to this Rip Van Winkle machine so we compromised by sowing half the field with oats and the rest with barley, and waited hopefully. Our optimism was sustained – quite wrongly – by the feeling that since a dreadful summer was dragging out haytime even into September, things were bound to get better for harvesting.

On the 18th of that month we had the thrill of seeing the first tied sheaves tossed out and the first circuit completed, with Fred driving the tractor. Giles sat behind, high up on his springy metal seat, operating the levers as instructed, putting the cutter in and out of gear and raising or lowering the height of the cut. It was most exciting. Fred hoped to complete the job in one or two afternoons. In the event three weeks were to pass before the "Deering of Chicago" – as proclaimed by a brass plate – could complete its work.

The 'Deering of Chicago' in action

## PROGRESS AND CHANGE

Most things one might imagine breaking down, did, and delay followed delay. A strap riveted to a canvas pulled out. Then the canvas itself tore badly. The two days lost in repairing these were beautifully fine. Rain returned with the repaired items. When business resumed the knotting mechanism failed so untied sheaves had to be recycled or tied by hand. Rain again! Then an essential small iron chock was jolted out and another day lost before finding it in the stubble. However every setback was useful education preparing us for future seasons when we should be on our own. Eventually the last oats were cut (September 30th) and the Deering trundled back home across the dale leaving us to stook the last few sheaves and begin to load the earlier ones onto a trailer heading for Black Carr buildings. That needed yet one more lesson. How to stack hundreds of slippery sheaves of oats on an open-sided trailer so they don't slither off on their juddering journey up the track is an art form of its own. All was safely gathered in by October 8th to our enormous satisfaction and relief. Subsequently we bought the Deering and used it happily for all the crops of oats we were to grow before we left Bank House.

Compared with these radical arable innovations our pattern of livestock changed comparatively little. Progress was represented by improved growth and health rather than greater numbers. Like most farmers we could not ignore the fashions and prejudices of the market but we held to our convictions about the excellence of the breeds we had chosen and limited changes to various crosses. Thus our Large Black sows were usually crossed with Large White boars except when we wanted to breed our own replacement sows. Similarly with our cattle the necessary periodic changes of bull were always with more Devons until 1987 when we introduced a cross with the French Salers breed with which the general public noticed no difference. The Salers shared very similar characteristics of temperament, easy calving and scavenging habits with the Devons: their main advantage was greater length of body and a higher proportion of lean meat to fat, which we certainly didn't want for ourselves but unfortunately much of the public did.

In the same way with our sheep we always retained a breeding basis of Kent ewes but crossed them with a variety of tups. Early on when buying in to make up numbers we soon learned to be wary of ewes with moor blood in them which they demonstrated not just by their ability to scale walls but by walking along the tops of them, preferring their own grazing pattern to ours.

In contrast to our breeding policies for four-legged creatures our acquisition of chickens was largely opportunistic, resulting in a delightfully colourful and delicately patterned flock.

To our foundational trio of escapist bantams were added Silver Wyandotte, Maran and Rhode Island, this last being a deliberate move to produce bigger table birds and lovely large brown eggs. Our annual egg production increased over the years until it topped 5000 which most egg producers would regard as mere peanuts but which we found highly satisfactory since financial profit was not our prime motive for keeping hens. Our gut instinct was that a farm should have hens and that these hens should range freely, cleaning the place up, putting a great deal of waste to good use and producing very rich manure. We were content as long as the cost of keeping them was entirely covered by the sale of eggs and birds, which it always was because the whole operation was carried out on a minimum outlay. In fact in addition to the benefit of all that we ate ourselves we reaped a cash profit rising from £24 a year to begin with to an average of over £200 for all our later years.

For us "free range" meant exactly that: we did far less to shut them in than to shut them out – from garden, from cars, and from the kitchen! Anyone coming through the farm on a dry day was bound to see chickens scattered everywhere near the farmyard, even way up the bank, and of course visiting the pig-feeding areas on the edge of the Nutwood. And yet they nearly always put themselves to bed in the chicken loft and the barn where we had put up perches. All we had to do was shut the doors. The hens chose where to lay their clutches of eggs and when to sit on them. However well hidden, we usually found them thanks to the hens' irresistible urge to trumpet the arrival of each egg to the world at large. Once discovered, the broody lady and her treasures were put in a portable hutch and taken to their nursery quarters among our fenced-off fruit bushes where the main flock couldn't raid either the fruit buds or the special chick crumbs. The fruit bushes throve wonderfully from the removal of weeds and pests and the liberal manuring.

A further practice of thrifty economy was to remove the maturing young cockerels each back end to the foldyards in Black Carr where they scavenged very productively among all the straw bedding put down daily for the cattle, and found the hay racks convenient, ready-made perches.

Our highly individual chicken system expressed our deep respect for the creatures, especially for the supreme maternalism of the hens, symbolised by their patient self-denial when broody, by their fierce protective aggression after hatching and by their habit, recognised in the Psalms, of spreading their wings out widely over their hidden families of bright-eyed chicks, which would suddenly pop out their heads through mum's feathers and just as quickly disappear again. What other farm animals show an equivalent to the skill and care with which a hen will teach its young where and what to eat even to the extent of pecking up a choice, single grain and putting it down in front of the chick?

The last addition made to our poultry enterprise arrived in 1983 in the form of eggs which we set underneath the maternal figure of Mrs Bond who didn't notice anything odd about her black family with webbed feet until they tried to take to water. They were Pennsylvanian Cayuga ducks which developed the most beautiful jet black plumage glinting alternately emerald and purple according to the angle of the sunlight. We kept Job, Kezia and Jemima to breed from and sold their siblings as table birds. Thereafter our account books recorded the sale of small numbers of ducks and their eggs along with those of the hens. They too ranged freely through the day finding a good proportion of their food for themselves and they performed a uniquely valuable service as expert hunter gatherers of slugs and liver fluke. They also endeared themselves to us by sharing a reluctance to quit their bed in the morning and by fulfilling the role of farm comics since we did not have any penguins.

The farmyard's first line of defence

# Chapter Twenty-one
# THE COMING OF AGE

Most of the progress and change related so far concerned our farming activity and showed the extent to which the early pattern of our work altered as the years went by. Whatever importance we placed on those agricultural details and policies, our first priority had always been the human element, and so it remained; yet for over half our time at Bank House we did not consciously depart from the largely passive way we acquired our helpers, or change our domestic arrangements as we lived with them cheek by jowl.

As with our resident helpers whose main purpose for coming was to work and gain experience, so with the other categories of visitors whether they came to recuperate, to sketch, to walk, to be stimulated or to see alternative farming in action. Having issued general invitations we left the initiative to them. The huge influx of visitors staying at least one night – fifty in our first five months – reduced after that first fine careless rapture; nevertheless the visitors' book continued to record a similar figure over each whole year. All of them carried away with them a much clearer idea of what we were about and nearly all were added to our Christmas letter list if not on it already.

The scale of our open hospitality was not without an element of self interest. It grew out of the basic purpose of our venture. It softened the social isolation we had plunged ourselves into, providing immediate congenial company while we were only beginning to establish friendships and social activities in our new community. This was of importance for relations between us and our young long-term volunteers, reducing the chances of our seeing too much of each other. Visitors made both working and leisure hours more varied and interesting: though they drew out mealtimes, imbuing them with a strong holiday mood, their efforts in house and field more than made up for it. Naturally, some were more rewarding than others. We benefited hugely from the advice of people of great practical experience such as Lady Eve Balfour (founder of the Soil Association) and St Barbe Baker (founder of the Men of the Trees). But everyone had something to contribute, from the octogenarian collecting twigs for the fire to the student keeping us up to date by explaining the latest "in" words . Even the drinkers and shopaholics proved useful when they discovered new sources of bargains for us and told us which pubs to patronise or to avoid. We couldn't be sure in advance which guests would prove most helpful. It was a fourteen year old who showed us how to re-

# THE COMING OF AGE

handle spades and forks cheaply, using sawn-off six-inch nails instead of expensive rivets. Another day three older lads were building a wall of concrete blocks. Two knowingly volunteered the third to drive yet another visitor to the station, five minutes away. He returned after ninety minutes having knocked up a neighbouring farmer to pull him out of a soft field with his tractor. He had mistaken a farm track for the county road. Meanwhile the other two got on like a house on fire.

Of all our visitors, one occupied a unique position and best illustrated Bank House hospitality in action; its most testing and most rewarding example. Mary's cousin, Richard, was born with cerebral palsy. Throughout his life he remained utterly dependent on full-time carers for all his physical needs. Unable even to crawl he could just shuffle crab-like over the floor where he spent most of his days. Most people found it difficult to understand his impaired speech. All his food had to be liquidised so he could suck it through a tube from a pint mug on the floor. A wheelchair was his only means of getting out of doors where he could be lifted into a car or onto a tractor-drawn trailer. Beginning in 1950 he and his full-time carer had enjoyed an annual holiday at Mary's family home in Sussex. Then after we got married he came to us at Millholme each year. Could we cope with him on the farm? By 1976, backed by helpers, we decided to give it a go; thereafter he came to Bank House for a couple of weeks just before haytime every summer but one until he died fourteen years later. For eight of his later visits we took over looking after him entirely, in order to give his carers a much needed break.

In spite of his condition Richard was no exception to our claim that every visitor had something different to contribute. We and our helpers, our neighbours and visitors, all wondered at his spirit, his patient and charitable disposition. He never grumbled or complained about his lot, nor while he was with us did anyone else, so benign was his influence. Deprived of any regular schooling, he had largely educated himself from the radio till he was widely knowledgeable, his mechanical and electrical understanding in particular being invaluable. Things that broke down during the year were set aside till he came so we could mend them under his instruction, acting as his hands. In this way our electric fencer was repaired. Mary used to leave the awful task of making VAT returns until Richard came with his calculator to make short work of them. It was also very helpful working out complicated seeding rates and acreages. We retrieved a cow, stuck belly deep in a quagmire, when he suggested we use our fore loader. He was able to sex our kittens when the vet lost his nerve and declined, because lying on the floor all day he was well placed to observe them closely as they played all over him. Richard always heard the weather forecast which we repeatedly missed and he rescued many a meal by pointing out to Mary that she hadn't shut the oven door. What we all owed to Richard was beyond compute and it seemed to us that with his advent our Bank House venture had come of age.

School parties accounted for many who came to look over the farm on day visits. A regular source of these was the County's Outdoor Education Centre at East Barnby, some six miles away. We soon came to recognise these uniformly anoraked children, spread sheets in hand, chattering their way down the bridle path on the bank, pausing occasionally for their teachers to point out pertinent features. When we discovered that the purpose of some of these outings was to teach them the difference between hill and lowland farming as part of their geography syllabus, we revealed that we were teachers in disguise. After that we began to take part, demonstrating how Bank House fitted the bill, even to the extent of Giles leading them up and down the steepest banks carrying posts or logs so that they really experienced that the term 'hill farm' means what it says. One teacher from Ripon found it so helpful meeting 'real live farmers' that she made arrangements to come several years running.

A nephew of Mary's who taught in Hull used us as a staging post for his pupils on their coast to coast walk, repaying us by spotting that our bees were swarming. Giles' niece, Katharine, created and ran the 30 acre City Farm on London's Isle of Dogs. She brought a car-full of cockney kids for a few days to see how different farming could be from their experience where they always had to have a paid night watchman on duty. Another relation, a young cousin on a teacher training course in York, had chosen agriculture as a theme for a project which was part of her Teaching Practice at a primary school. Over thirty youngsters, packed into a fleet of parents' cars, arrived in a state of high excitement ready for the programme we had devised for them. We still cherish the sheaf of brightly -coloured thank -you letters that they wrote when back at school, containing the following personal comments amongst many others:

'The cow was lovely. I couldn't get my hand round her teat.'

'I really enjoyed making butter.'

'The best part was feeding the sheep because they tickled your hand when they ate the pellets up.'

'On the moors it must be very cold in winter.'

# THE COMING OF AGE

'I thought the squelching in the mud was best.'

'The best bit was when the horse chased the pigs and the pigs chased the cows and the bull just stood there.'

From the other end of school days a group of post-exam sixth-formers came to help trace and excavate some very old, blocked stone drains; an operation combining farming with archaeology. The value of all these visits depended very much on the preparation by the teacher who brought them. Once a small gang of teenagers were camping on their own unsupervised. We were surprised when they knocked at our door late on their first evening. 'Please can we use your toilet?' Obviously they hadn't been taught much about camping!

Our educating tradition survives us in various ways. Each year children from Glaisdale play group are still taken round the farm on an open trailer and our successors usually accede to requests to camp, directing applicants to the far end of the farm where they set up their tents beside the beck, a stone's throw away from the waterfall. If a visitors' book were still kept it would not be much less full of names than it was in our days.

Complementary to all these contacts with the outside world which brought people onto the farm were the occasions on which one of us was invited to address people elsewhere. It started with Mary giving talks to Women's Institutes. Then began a series of trips back to Abbotsholme each January with Giles in the role of a visiting preacher so as to justify the school's paying our costs. The necessary so-called "sermons" drew heavily and vividly on our farming experiences; the innocence of lambs; the maternal protectiveness of hens; the gold of corn in good measure pressed down and running over; the nature and value of hard work. Once we were asked to address a Quaker weekend conference in Ilkley on the subject of "simplicity in living". It was recorded briefly in their journal "The Friend".

Such developments were our response to initiatives on the part of others not changes made by us. Our attitude to visitors to Bank House stayed as welcoming as ever and our later Christmas Letters contained such remarks as: "Vintage year for visitors, so our guests' beds remained well aired." In sharp contrast it was the situation regarding our long-term resident helpers that was to change radically during our second decade, not in an attempt to invent a better system but as unavoidable adjustments to new circumstances, both internal and external. On the one hand it might have been expected that life would have become easier as we moved into the 1980's for we had broken the back of restoration. On the other hand we were not getting any younger and couldn't maintain our previous level of exertion indefinitely. Since these factors balanced each other it was the external changes over which we had no control that obliged us to reconsider the basic principle of a workforce of volunteers. The relevant background was the politics of the Thatcher government and the issues she had to deal with: unemployment, trade unionism and the Common Market Agricultural Policy. This last supported family farms which were mostly very small and which Thatcher's city-biased viewpoint dismissed as inefficient by criteria that left social costs and values largely out of the equation. All this seemed to have an unfortunate effect on our supply of volunteers. We began to find fewer of the young disposed to engage themselves in unpaid or poorly paid work. The carefree confidence in the future shown by our earlier helpers became harder to find.

We realised that two other considerations made us a less appealing prospect: our very success in improving the farm meant that we no longer appealed as a rescue operation on a tiny budget; nor were we as rare a chance to farm organically. Oddly enough, our having periodically increased the weekly pocket money allowance from the original £3 in 1973 to £10 in the early 80's as it became possible, made it a monetary calculation instead of an act of sacrificial generosity. A further deterrent was created by the new regulations governing unemployment benefit which now penalised anyone moving either way between employment and unemployment. This made it a much riskier business coming to us for a trial period. Yet another social change was working against us. Lack of money hadn't prevented our helpers in the 1970's from travelling freely and free, so widespread was the practice and acceptance of hitch-hiking. It seemed symbolic of the Thatcher era that drivers began to regard hitchers as scroungers if not as people to be feared. The change was neither sudden nor ubiquitous, but it was distinct enough to affect potential volunteers most unfortunately, making them feel that they would be more isolated.

Whatever the reasons for the difference, in 1979 we were writing that "our cruse of volunteers is apparently bottomless", whereas our letters in the early 1980's were full of disappointments and difficulties. Bad luck played a part – a girl who agreed to come for a year, cried off having broken a leg at home. A young man who wanting to work with horses, arranged to be met at the station on a Friday but he just wasn't there. When Mary phoned his landlady she said he had left on the Thursday to come here. We never had a word. A girl of student age arrived intending a long stay: overlapping with her was a lad preparing to be a farmer. Both had enjoyed a week's trial period earlier and seemed capable and willing to hold the fort while we

# THE COMING OF AGE

spent two nights away. She walked out the day after we returned, while he told us he had decided he now wanted to be an accountant not a farmer. He too left after we had talked it over with his sensible parents, so we faced the coming winter on our own. In fact we did get some helpful replacements, but with such worries becoming increasingly common we turned our minds more seriously to alternative solutions.

Two factors limited the field of people we might attract to join us: the lack of pay and the lack of separate accommodation. Originally we could do nothing about either but after ten years of improvement we began to reconsider both the inevitability and the desirability of our domestic pattern. A young couple in their early twenties turned up attracted by what they had heard of our way of farming, keen to learn from us and to join us. They were, to resort to modern jargon, in a "stable relationship" and wanted to settle down. With free board and keep they were prepared for both of them to work for weekly pocket money of only £15 at the time when our usual rate had become £10 for each helper. They would mix in with everyone else as far as catering and eating were concerned, they just needed their own separate room. This provoked us into taking up the long dormant idea that we should create a separate dwelling by converting some of the farm buildings across the yard. Not wanting to wait, they moved into a windowless loft reached only by a wide, loose set of wooden steps, and we started making detailed plans for a very small second home. All began famously until tragedy struck in the form of untreatable hay fever. After struggling valiantly but in vain for two months, they had to abandon the scheme altogether and departed.

That was a sad setback for all concerned: it was not however a case of back to square one. We had not just been building castles in the air, but detailed, practical plans and we had even involved a village builder. Metaphorically we had laid the foundation stone of the future "Toft". Meanwhile our need for help was answered by an unforeseen temporary solution. At the Whitby Folk Club we had made friends with Joe and Agnes, who were living on unemployment benefit with their young son. Joe had suffered being out of work for four years without any creative outlet for his great strength and energy. To dissipate his frustration he rambled solo all over the moors, studying and photographing the rich legacy of standing stones and other relics of antiquity. He loved helping on the farm occasionally but we couldn't pay him for his work without his having it docked from his unemployment benefit. At that time when we were short of resident volunteers Joe's support was invaluable. We found a variety of ways of rewarding him legally, though inadequately, such as giving him driving

## THE COMING OF AGE

lessons in our car so that he passed his driving test. In addition to the ewers of beer with which we celebrated the ends of most long haymaking days we fed him generously when he was with us and sent him home with presents of farm and garden produce. He was also able to gather wild fruit with which to produce large quantities of excellent home made wine.

It is important not to exaggerate the difficulties of these years. There were very few times when we carried the farm on our own and they were brief and enjoyable. What is more they witnessed a great deal of the progress already described. A combination of other factors distinguished our later from our earlier years apart from the capricious supply of volunteers. As we explained in our Christmas letter of 1984:

*"When we first came to Glaisdale we found our work and surroundings so absorbing that international events lost much of their significance. The connections between our animals and the world's food supply or between our volunteer helpers and widespread unemployment, were hard to discern and we were too busy to discuss them.*

*Much has altered in a decade. Perhaps the main reason why the miners' strike occupies our minds more than the chronic ill-health of a tup or the service of a sow is that we spend more time watching Panorama and the News or reading the papers."*

The outer world was becoming hard to ignore. Farmers devoting themselves to the production of food were peculiarly vulnerable to appeals for help for the starving of Ethiopia or the victims of Bhopal. (A neighbour harder up than we were instantly promised a calf.) Besides beginning to have more time we were finding our own financial situation easier, helped by the trickle of customers who thought it worthwhile coming to the farm in person in their search for food free from "badditives" and paying more for our woodland pork. Public opinion, and even the government, were beginning to notice the "green" movement and we were no longer feeling that we were an oasis of irrelevance. Nevertheless we knew we should sleep more soundly once we were no longer living hand to mouth as regards our workforce. It was a form of excitement we had once relished but could now do without. We never lost our conviction that something or someone would turn up. Sure enough it did. "Could we write a reference to support an application for a job?" We certainly could.

The request came from Chris Padmore who as a helper some seven years previously had shown an instinctive appreciation of our way of living and working. Despite having a degree in Agricultural Science he had been unemployed before going to London where he was driving a delivery van. Now he was trying to get back to farming. By an extraordinary coincidence of which he was quite unaware the post he was applying for was for someone to succeed Giles' niece Katharine who had decided it was time to return to full time architecture and let her City Farm stand on its own feet. That grapevine told us that the wrong person was appointed. So here we were wondering whether to risk employing a full time assistant (the main risk being compatibility rather than financial) and knowing that there was Chris wanting to farm again. It was February 1985. He came for a week's refresher course to ensure that there was more in it than nostalgia. Finding that there was he agreed to return in time for haymaking. On our part we had to do a lot of sums, not just to know what salary we could guarantee but also to blow the dust off our conversion plans to provide him with a separate dwelling which we soon baptised with the medieval term of 'The Toft.'

Chris would live with us to begin with on the understanding that he would be able to enjoy his separate bachelor's pad as soon as possible. Sad to record, on this last point we failed him grievously. Month after month slipped by without any visible work on the conversion. In the end it was not until we came to write our Christmas letter for 1987 that we were able to report that:

*"This has been the year of the Toft when the dream of a decade has been realised and an ideal home created out of some old farm buildings. Now at last Chris is master of the Toft."*

Apart from the delay, we were inordinately proud of it as a work of art and craftsmanship. It was our most important achievement at Bank House since the complex of buildings in Black Carr over ten years earlier. Besides its enormous usefulness to the working farm we saw it as a contribution to the wider community for which we were adding much-needed housing which was in due course to send four young children to the village school. Its construction provided valuable local employment, all the skilled work being done by men living in Glaisdale or by their regular collaborators. In many respects, such as its wood-burning stove for cooking and central heating, the Toft was all of a piece with the life of thrift we have outlined. Considering the quality of materials and craftsmanship the total cost was remarkably low, thanks in good part to the amount of donkey work done by the home team, and to the availability of redundant stone gateposts for conversion into lintels, cills and doorsteps.

We assumed it would be the one chance of our lifetimes to design a house and we wanted to make the most of it, which explains why we couldn't forebear to sing its praises in that year's Christmas letter. Having described the kitchen's red-tiled floor, gentle blue worktops, white walls and window seats it waxed eloquent about the upstairs rooms.

> "The sensuous curving plaster in the apex of the roof sets off the strength and dignity of ancient, unvarnished trusses, while the angles left by the dormer windows, changing floor levels and the sidestepping chimney, create an intriguing space that in plainer dimensions might have yielded a mere box."

The source of inspiration was not far to seek as the same letter acknowledged.

> "Many of us contributed ideas but it was Katharine's presiding genius that welded domestic comfort with visual beauty without losing touch with farming ancestry. Every farmer should have an architect niece."

Much as we enjoyed the Toft architecturally its true importance to us lay in the fundamental change it enabled us to make in our farming. It provided permanent accommodation for a mature assistant, someone of experience capable of releasing us from nearly continuous presence, responsibility and physical work – we had had to wait seven years for our first break of more than a weekend. It also released us from the recurrent necessity of training raw recruits in every aspect of daily routine. We could, and did, continue to welcome visitors wanting to help, in whose supervision Chris took a big share. These always included a notable proportion of past volunteers who would settle instantly into whatever work was in hand. Fortunately we began to experience the benefits of the new regime as soon as Chris rejoined us though unfairly he had to wait before he reaped his due reward from the bargain.

The benefits of the new regime were many and various, ranging from the ability to tackle long-postponed improvement projects on the farm to the opportunity to follow and participate in movements of public interest in the outside world, locally and nationally. In between we were increasingly able to join in all sorts of social activities in the community around us. The best example of improvement was the restoration of the track rising some 250 feet up the bank from the farmyard. As we inherited it, it was hard to credit that until 1947 it had been what passed in those days as a council road, linking the south-eastern side of Glaisdale dale with the rest of Egton parish to which it belonged. However a spade or a crowbar thrust into the blanket of peaty soil and rush roots confirmed that some nine inches down there was still a continuous layer of broken stone, in fact the un-tarred macadam

# THE COMING OF AGE

road of olden times. That hardest of winters had set off a landslide at the steepest point which carried away over half the width of the road, leaving a narrow passageway for animal and human feet alone. Our interest in the track was not that it had led to Egton Bridge but that it had been the only direct route for wheeled vehicles to reach our three top fields bordering on the moor, fields on which we were told potatoes used to be grown. Our ambition was much more modest: to remove the cover of rampant bracken, thistles and rushes and restore those seven acres to useful grazing.

Apart from having had lime and basic slag spread on the one field deemed to be accessible by the contractor – who had flattened yards of boundary wall to gain that access – our only work of restoration so far had consisted in "topping" the fields annually. Though the nearest of these contiguous fields was within shouting or catapult distance of the farm house, getting a tractor and cutter-bar there involved driving seven miles round by road, then the hair-raising mowing itself, wondering all the time how near to the unfenced precipice it was safe to mow, and then the relieved seven miles home again. It is not surprising that the lure of making the track usable by tractors was irresistible. The crux of the task was the site of the landslide itself. To treble the width of the remaining track we had to hack back the rocks holding up the hill above. Pick-axes, crow-bars, shovels, sledgehammers, cold chisels and wedges were all needed, but we succeeded, and having made sure there was a way through for tractors getting that far we began digging out the silt of approaching forty years starting at the bottom.

It was one of those endless, insatiable jobs, swallowing up time and energy whenever the rest of essential farm work allowed. Many unsuspecting visitors found themselves putting in a stint of elementary navvy work for little skill was required, but its saving grace was that the view got better and better with each yard of progress until near the top a saucer of sea came into view between the cliffs above Whitby.

Before long all that invested labour paid off handsomely. Next to the eastern-most field, just over our boundary wall, a very large area of self-sown softwood trees had grown up. At this time it was clean felled preparatory to systematic replanting. Their attempt to burn a vast amount of wood ended in disaster with a fire engine in attendance. We offered to lessen their problem and were given permission to remove anything we wanted for firewood. As with digging out the track up the bank we now had an endless, unskilled job on our hands; another trap for visitors. How much free firewood did we need? How long is a piece of string ? For months we lugged trunks and branches, great and small over a little stream and into our nearest field where we amassed great piles, enough winter fuel we hoped for four winters or so. Thanks to the newly re-opened track it could wait there indefinitely for us to ferry it down to the farmyard in our own good time. This we did with a tractor carrying a buck-rake which was so heavy, and the track so steep in places, that going uphill even though unloaded, we had to have someone perched on the front of the bonnet to keep the front wheels down on the ground.

Six years before this success we had embarked on the restoration of the gently undulating track through the Nutwood. Part of the original cart track linking the two small farms that comprise the present Bank House holding, it had degenerated into a slithery bridleway that required more than walking shoes for much of the year as it rarely dried out even in summer. The work consisted chiefly of unblocking or rebuilding many essential little culverts that channelled a series of small streams under the track, and of digging short stretches of gutter on the upper side to divert water into those culverts. It was an added incentive to restore this track that under our Farm Scheme contracts with the National Park we were to coppice small areas of the Nutwood each year. Compared with the obvious usefulness of the bank track, that in the Nutwood lacked urgency so the work was only spasmodic and progress slow. Nevertheless we did eventually complete the task. If we had possessed enough flags there would have been a lot of flag waving the first day we drove a tractor right through the Nutwood and emerged on the other side. Another historic landmark had been reached, though its value was symbolic as much as practical.

The period of our partnership with Chris while he was still a bachelor was a very happy and productive one, allowing us a new, unfamiliar, freedom to get away from the farm. This was just as well because in 1986, after her mother's death, Mary's duties as chief executor necessitated eight separate trips to Sussex. Her absences, which added up to 39 days, must have contributed to the disappointing delays in preparing the Toft. We made the most of our new freedom in the next couple of years: besides going to weddings in Oxford, Hull and Kendal, the reception for the last of which was reached by the Gondola across Coniston Water, we had a rare holiday near the Kerry border of West Cork, visited friends running the school for maladjusted children that they had founded north of Inverness and sang once more at the Music Camp reunion in London. Nor was this all because to add to these joint excursions Mary attended Liberal Democrat annual conferences and Giles escaped to France for a brief fortieth anniversary

## THE COMING OF AGE

reunion beyond Versailles where he had spent a year helping the French Red Cross caring for children mutilated during the war.

As our social life took us far afield, it also developed fruitfully at home. Mary's occasional choir, specialising in unaccompanied singing, was born in Bank House kitchen in 1978 and still flourishes. Its annual Carol Service has long been a welcome ingredient of the Glaisdale Christmas season, as have been the Beggar's Bridge Players' pantomimes for which Chairman Giles rang the changes as actor, author and producer. It was to be one of the more improbable turns of events shaping our future life in Glaisdale that many of the people who elected Giles as their Borough and County Councillor knew him best as a pantomime dame.

Some of our freed energies found more serious outlets. These arose from our deep concern about the related threats to dales farming and to rural communities as a whole, so when in 1983 we heard that a Small Farmers' Association had been established to campaign for the survival of small family farms, we signed on. Mary became a very active member of the SFA, (later the Family Farmers' Association), lobbying on its behalf as the main focus of her Liberal Democrat activity, so that years after we ceased actively farming ourselves she was being invited by MAFF and then by DEFRA to represent the FFA at consultative conferences. It was largely owing to Chris that the farm continued to thrive while we took part in public affairs for more and more of our time. It seemed too good to be true, but could it last?

That question was about to be answered by the kind of fairy-tale ending that would not have been out of place in one of our Glaisdale pantomimes. Chris loved Bank House and was in no hurry to go anywhere else, but with characteristic directness and honesty he let us understand that some day he wanted married life and didn't see much likelihood of its coming his way while he was so isolated and so fully occupied as he was here.

Our story at Bank House had already contained turns of almost unbelievable good fortune, more in keeping with fairy tales than with sober history, as when we found and bought the farm. And surely a fairy godmother was waving her wand in 1986 when through a series of coincidences and with an eye on distant retirement we bought The Cottage cheaply at the only time in our lives that we had the ready money needed . We had often wondered how we could ever bear to leave Bank House with its wonderful position overlooking so much of our lives but then we hadn't dared to hope for such a sequel, still in sight of the farm and on the sunny side of the dale. We were not alone in regarding that house and garden as the most desirable in Glaisdale yet we secured it without the neighbourhood

# THE COMING OF AGE

even knowing it was for sale. We set about letting it out for rent and carried on farming with Chris for four years until in 1989 when the wand was at it again. To recapture the freshness and excitement of the time we shall relate what happened next in the words of our Christmas letter that December, but first we should explain that Emma and her twin sister had been third year pupils at Abbotsholme the year we left, and had taken up our general invitation to Bank House, by persuading their mother to drive them over to the farm for a day. Emma had never forgotten it:

*"Some of the biggest question marks that hung over the future of Bank House last Christmas hang no more. What has given the intervening year an aura of make-believe has been the speed and certainty of important events, the way the jigsaw puzzle suddenly slotted into place. Barely a twelve month back we answered the knock at our door of the sales manager of a bandsaw company and in walked Abbotsholmian Emma with her Giaconda smile, following a powerful hunch that this was the place for her. A second visit brought some gumboots which got left behind. They are still here, and so is she. Skimming the ensuing pages of our story you may light on a September day of unrivalled sunshine, the perfect setting for the wedding of Chris and Emma, the fulfilment of the Toft. Weddings begin more than they end and in our Christmas novel this wedding plunges the reader into the second volume of the Bank House saga wherein the Heron-Barran era passes gently into history as the Padmore-Vodden one takes shape. What makes the present so exciting and rewarding is that far from any sort of vacuum or interregnum there is a kind of dual monarchy. We are enjoying a period of overlapping energies, a doubling up of resources and support, the cross-fertilisation of new ideas with tried experience, and this will be greatly reinforced when Emma gives up her outside job at Christmas."*

*'Retirement' is inadequate to describe the biggest change in the second half of one's life, obscuring its character and belittling its importance. For many the experience is instant and total. Whether welcomed as release or dreaded like leprosy it is a cruel shock to the human system, a procrustean procedure. Why not go gently into that good evening of declining strength? It seems more natural, more in keeping with the organic way of life, and we are fortunate that we are able to set out on such a path".*

which is why as late as the end of 1991 we were writing:

*"Everyone asks us if we have retired but it is not easy to give a precise answer. 'Yes and No' may not be very helpful but it comes pretty near the mark."*

# THE COMING OF AGE

For example looked at from the point of view of who was running the farm, there was no one moment of transition. Since we had already passed the moment of recognising that the most hopeful solution for the future of our Bank House project was for Chris and Emma to run it, every important decision was shaped to that end; everything was discussed between us, and who could say which of us had first suggested what we agreed to do? Who cared anyway?

It was our foremost desire that what we had created should not be dismantled as soon as we were no longer running it, but should continue as far as common sense and changing circumstances allowed. Goodbye to that if we just put up the farm for sale. We knew only too well that we should never have survived financially farming as we did if we had had rent and a mortgage round our necks. Nor could we expect Chris to do so even if he could raise a mortgage which his years of unemployment and low pay made an unrealistic proposition. Fortunately we didn't have to sell the farm or charge a rent. Our guardian angels had seen to it that we already owned our future home and we were not tempted to abandon our practice of thrift as a way of life. So, secure on that front, Chris and Emma could concentrate on earning enough to live on from their farming, in much the same way as we had. We recognised that similarity in our Christmas letter of 1990:

> *"The economic brink on which Chris and Emma now stand is every bit as daunting as what we faced in our penniless days when rescuing the farm from dereliction. They may begin with fertile land, fine stock and a Green tide but the predicament of agriculture is dire. This year our chief source of income was halved by mad cow disease, the hideous offspring of efficient greed."*

That was the year our residence at Bank House drew to its appropriate end, hastened a little but very willingly by the arrival of Joseph, his early cries mingling with those of lambs and calves of his own age. We had designed the Toft for a bachelor, or hopefully, for a young couple, but it was not an adequate home for bringing up children. With The Cottage available and with Mary approaching 68, what was there to wait for? All the time The Cottage was let out she had been busy transforming the garden there, her dedicated, creative work for the next stage of our lives. The four of us agreed that we should move to The Cottage at the end of November leaving Chris and Emma in the Toft to supervise important improvements to the Spartan farmhouse while it was uninhabited. Farewell to the concrete floor in the kitchen! Welcome the new Aga and some central heating! For the time being the partnership would continue with Giles working on the farm full time – insofar as it could be full time for someone not sleeping on the spot in lambing time.

Despite the adverse tide of agriculture in which they operated Chris and Emma passed their first year at the helm with flying colours: may the reader pardon yet one more, last, quotation from a Christmas letter in order to diminish the risk of rosy tinted spectacles:

*"They have crammed their first year as only youthful energy can, establishing an outlet for the sale of our livestock through the Pure Meat Company, qualifying for full symbol status with the Soil Association for our sheep and cattle, negotiating a second and extended contract with the National Park which will yield grants in support of our conservation practices and undertaking an overlapping land improvement scheme with ADAS.*

*Progress at Bank House contrasts with the plight of many local farms. Some farmers have had to sell up; others hang on from loyalty to the past rather than from hopeful expectation of tomorrow and many of their sons are retreating from the land. For farming is being rendered insecure by political vacillation, hamstrung by ever-changing regulations and restrictions, crushed by mortgages, caught in the crossfire of the public debate on the environment and left exposed to fight the more prosperous sections of the community for houses and land. It is a happy irony for us that after years as the self-appointed Cinderella among the farms of the dale, Bank House stands a healthy chance of survival now the dangers are much greater."*

Mary and Giles at The Cottage

We too have crammed our lives with activity and interest. The distant image of retirement when one can read a novel in an easy chair on a weekday morning proved a chimera, fortunately! But there are moments when we can stand and stare. From The Cottage windows and garden much of the farm is clearly visible so it is never absent from our thoughts for long though we go there less and less. The nearest point of it is the still largely untamed, sheltered area of bracken, crab apple and other trees which we call Arnecliff End. From here the ravine is out of sight as are the daffodils above it. The things the eyes cannot see are nevertheless deeply etched in our minds; we know they are there; the vinegar stone just there; and the waterfall below about there, and the bluebells where a song was written to the left over there. Every few yards evokes an incident; the helpers fencing in the snow all along that stretch, the children playing in that stream and the red cows hurrying up that slope for remembered windfall apples, those beautiful red Devons that underpinned our livelihood for eighteen years, not a herd but a family of individuals known by name. Let us end with a particular memory on a particular morning with a particular cow in that particular patch of Arnecliff End before we had fenced it in. It was Sunday 20th June in 1976 and was recorded in a letter written in precious weekend time off:

> *"Plum, the tamest cow of all, nearest to a pet, has got a lovely daughter. She and a select band were put through into Arnecliff End last Monday to eat the lush grass among the bracken and bluebells and remains of daffodils: 'select' in the hope that these particular ones wouldn't roam and take advantage of there being no fences to prevent their going to sea! It has been a game trying to find the calf among twelve acres of wood and bracken each day. Plum and the calf hide. I find Plum. Plum says her daughter isn't at home. I try to interest myself in local varieties of undergrowth. Plum looks into the distance. I try various sorties round the compass until I hit on the right direction, when she follows me in case I find the treasure. I say thank you for the hint and begin another round of exploratory walks, watching Plum for an indication whether getting "warmer" or not. Eventually I see a deep red patch through the bracken stalks, and a pair of large, unblinking eyes. Then we agree all round that life is good, and part till tomorrow."*

On the edge of the ravine

# Acknowledgements

The gestation and birth of this book involved so many people over so long a period that I cannot give due acknowledgement of each person's help. For example, of the splendid collection of photographs amassed since 1972, we have no complete record of who took what. I apologise therefore for failing to fulfil the normal courtesy of expressing thanks for a great many individual contributions, whether photos or anecdotes, constructive comments on draft chapters, or the correction of factual errors.

It was continually encouraging to meet with interest and appreciation at every stage of the work, support that was all the more valuable when progress threatened to grind to a halt. We were sustained by other people's conviction that it was important for the *Bank House Story* to be told, even at the eleventh hour.

I welcome this opportunity to record my appreciation and thanks to the following people for their help in very diverse ways. Collectively, besides solving problems great and small, they made the production of this book an enjoyable and fruitful experience:

Chris Padmore ; Joanna and Alan Gent; Celia and Edward James; Dom Benedict Heron; Garth and Delia Haythornthwaite; Irene and the late Ray Proud; Nick Wates; Oliver Wates; Jeff Balls; Philip Conford.

It should not be taken as any diminishment of my gratitude to the above that I single out yet others without whose major help the book would not exist. Their months of dedicated and meticulous work transformed my sheets of scribble and flow of ideas into this handsome volume and ensured its physical quality.

Emma Padmore performed the unenviable and seemingly endless task of typing the original draft text from my very amateur use of a dictaphone as chapter succeeded chapter.

The map and black and white drawings are the handiwork of artist David Moss whose intuitive visual understanding of Bank House was nurtured by fourteen years' residence in The Toft.

Mike Lyth employed his impressive professional grasp of the technical skills needed in book production. As teacher, and now friend, he needed all his patience to modify our wilder aesthetic ambitions and inconsistencies while interpreting the business realities of our co-operation with Authors OnLine.

And then there is Mary. Since it was left to me to shape each chapter and wrestle with the form of words, the reader may well under-estimate Mary's enormous contribution to the book that rightly bears her name. She has done everything but write it. Every page was submitted to her scrutiny and greater knowledge and was discussed in detail. She spotted innumerable errors and weaknesses and left her stamp on subsequent improvements. It is in every sense, therefore, not 'my' book but 'our' book.

Printed in the United Kingdom by
Lightning Source UK Ltd., Milton Keynes
142556UK00001B/130/P